Safety Culture and High-Risk Environments

A Leadership Perspective

Sustainable Improvements in Environment Safety and Health

Series Editor

Frances Alston

Lean Implementation: Applications and Hidden Costs, *Frances Alston* [2017]

Safety Culture and High-Risk Environments: A Leadership Perspective, *Cindy L. Caldwell* [2017]

The Legal Aspects of Industrial Hygiene and Safety, *Kurt W. Dreger* [2018]

Industrial Hygiene: Improving Worker Health through an Operational Risk Approach, *Willie Piispanen, Emily J. Millikin, and Frances Alston* [2018]

Safety Culture and High-Risk Environments

A Leadership Perspective

By

Cindy L. Caldwell

CRC Press is an imprint of the
Taylor & Francis Group, an **informa** business

CRC Press
Taylor & Francis Group
6000 Broken Sound Parkway NW, Suite 300
Boca Raton, FL 33487-2742

CRC Press is an imprint of Taylor & Francis Group, an Informa business

Printed on acid-free paper

International Standard Book Number-13: 978-1-138-03506-5 (Paperback)
International Standard Book Number-13: 978-1-138-10524-9 (Hardback)

Visit the Taylor & Francis Web site at
http://www.taylorandfrancis.com

and the CRC Press Web site at
http://www.crcpress.com

Dedication

To Brigadier General Roswell E. Round, Jr., a truly authentic leader, husband, father, and friend.

Contents

SECTION II

Preface

The complexity of advanced technologies and organizational communications in an outsourced global business environment has increased the risk of devastating accidents. Risky processes rely on people to understand the complex relationships associated with the systems they operate. Errors connected to human behaviors are a by-product of the organizational culture. The literature suggests that organizations using high-risk technologies may become more resilient by adopting a strong safety culture and high reliability principles. Leadership is widely recognized as playing an important role in influencing safety culture and creating more reliable organizations.

This book is written for the leader who manages the risks of hazardous processes. It reflects on the role of authentic leadership in influencing safety culture and organizational resilience to failure and provides a practical guide for leaders to assess, monitor, and improve their culture and capacity for resilience. The general principles and tools are applicable to all high-risk domains from banking to healthcare to nuclear power.

Author

Cindy L. Caldwell is currently a senior technical advisor for Environment, Health, and Safety at a Department of Energy National Laboratory. Her work includes understanding and evaluating operational culture, organizational reliability, and risk management. Ms. Caldwell is certified by the American Board of Health Physics, has a BS in bacteriology, an MS in radiological science, and a PhD in human and organizational systems.

The materials and discussion provided are her own and are not the opinions of her employer or sponsor.

Section I

1 Understanding Organizational Factors in High-Risk Environments

Historically, accidents within an organization have been explained in terms of technology and human factors without taking into consideration the organization's values, beliefs, and behaviors. These cultural dimensions can also create limitations in organizational systems. In certain sectors, such as nuclear power and petro-chemical industries, the complexity of the organization–technology interface introduces unforeseen interactions among system components that have resulted in catastrophic accidents. High-risk industries such as the nuclear power industry became interested in culture in the early 1980s. The International Nuclear Safety Advisory Group (INSAG) first coined the term *safety culture* when referring to the failure at the Chernobyl Nuclear Power station (Sorensen, 2002). INSAG (1991) acknowledged that after a certain point in the maturation of safety systems, technology alone cannot achieve further improvements in safety; instead, organizational and cultural factors become more important.

1.1 SIMPLE TO COMPLEX

The early evolution of the management of safety-related programs in the United States began with a compliance-based inspection focus brought on by the Occupational Safety and Health Act (OSHA) in the 1970s. As a result, procedures, training, and tools were established and continuously improved to increase awareness, reduce risk, and ensure compliance. Although OSHA and its enforcement powers had a positive effect on safety in the United States, reducing fatalities and disabling injuries by more than 50%, there was limited attention to long-term performance or continuous improvement.

Likewise, for most of the twentieth century, accident theories were traditionally based on a closed-system approach that examined conditions, barriers, and linear causal chains that do not consider the complexity of today's working environment. The typical focus in the 1960s and 1970s was on technical faults and human error, which received international attention following the accident at Three Mile Island.

The accident at Three Mile Island was compounded when plant operators failed to recognize the warning indications of a loss-of-coolant accident due to inadequate training and human factors. In particular, a hidden indicator light led to an operator manually overriding the automatic emergency cooling system of the reactor because the operator mistakenly believed that there was too much coolant water present in the reactor (Health and Safety Laboratory, 2006). The Three Mile Island accident investigation's Kemeny Commission coined the phrase "operator error," which reflected the cause and effect thinking of the 1970s (Kemeny, 1979).

In a similar fashion, the compliance-based "workplace inspection" approach of the 1970s gave way to behavior-based safety in the 1980s and early 1990s. Behavior-based safety was grounded in the belief that individual behavior could be changed by focusing on the behavior itself and the consequence of that behavior. The approach reinforced the idea that an unsafe behavior was the result of individual choice without regard for other organizational factors. Hollnagel (2009) has noted that many accident models still share a linear approach and are silent on the dynamic interplay among factors. Linear causality suggests that the effect is proportional to the cause. Causal explanations of incidents provide greater organizational control by placing blame on the specific individual without considering systemic issues behind the incidents. The organization socially constructs a view that the essence of safety is to prevent individuals from committing errors (Reiman and Rollenhagen, 2010).

Unfortunately, managing worker safety by regulatory compliance and addressing individual error does not address the risk of a catastrophic event. Catastrophic risks need to be treated differently from other risks. Executives can prevent catastrophic events through a holistic organizational focus on risk. Low-probability, high-consequence occurrences that have a high inherent risk such as Three Mile Island are called catastrophic events (Kleindorfer et al., 2012). Historically, catastrophic events have played an important role in the development and application of accident theory, so it is not surprising that an alternative view emerged from the Three Mile Island crisis.

Sociologist Charles Perrow disagreed with the human factors argument presented by the Kemeny Commission. Perrow (1981) proposed a sociological view of the relationship between reactor operators, their indications, and decision making under duress. Perrow (1984) explicitly stated that it was the system that caused the Three Mile Island accident, not the reactor operators. Perrow proposed that an association should be made between the errors and the system, as opposed to the errors and the operators. Perrow's 1981 paper included a description of a normal accident that later became normal accident theory. Normal accident theory is based on the unanticipated interaction of multiple failures. Perrow (1984) believed that if the system is both interactively complex and tightly coupled, there is no possibility of identifying unexpected events and the system should be abandoned.

Perrow's arguments aligned with limiting the human interface and were later reflected in a paper by Otway and Misenta (1980) that concluded by questioning whether the role of operator should be eliminated or redefined for less human intervention in emergencies. It was the safety community's negative reaction to Perrow's (1984) controversial ideas that led to much of the present-day thought on safety culture. However, it wasn't until many years after the Three Mile Island event that culture took on a more prominent role in the prevention and analysis of accidents.

The concept of organizational culture became popularized in the 1980s with bestseller books such as *In Search of Excellence* (Peters and Waterman, 1982) and *Corporate Cultures* (Deal and Kennedy, 1982). Emphasis was placed on the importance of employees as a resource and management as an influence on cultural change. Since the 1980s, the concept of organizational culture has been extensively studied and has become an interdisciplinary interest spanning the fields of psychology, sociology, and management. Culture is shared among groups of individuals and is not a characteristic of single individuals. It is something that members of an organization learn over time as the correct or wrong way of behaving in an organization. Schein (2004) describes culture in terms of three elements:

- Culture offers structural stability that provides meaning and predictability to the organization. Culture survives even when some members of the organization leave. Strong cultures put considerable pressure on people to conform.
- Culture often resides in the deepest subconscious part of a group.
- Culture influences all aspects of how an organization deals with its operations.

Using organizational culture as an instrument of control remains a source of debate and today many call for a more holistic approach to understanding culture that considers its many layers and complexity (Haukelid, 2008). The early concept of safety culture emerged from the popular view of organizational culture and it faces similar debates. Cooper (2000) noted that safety culture does not operate independently, but rather it is tied to other non-safety-related operational processes or organizational systems.

Since Chernobyl, numerous definitions of safety culture have appeared in the literature. The preponderance of definitions refer to culture in terms of attitudes and behaviors (Guldenmund, 2000; Hale, 2000; Lee, 1996; Ostrom et al., 1993; Pidgeon, 1991). Most definitions have focused on safety culture in high-risk areas such as nuclear power, mining, and transportation industries (Guldenmund, 2007, Wiegmann et al., 2001). The majority of definitions suggest that safety culture is a discrete entity that an organization "has" and that culture can be understood in terms of perceptions and workplace behaviors. These definitions imply that culture can be changed.

The interest in safety culture stimulated related research in accident management. In 1985, a group of researchers at the University of California, Berkeley, initiated research on high reliability organizations (HROs) by examining organizations that operated virtually error-free, such as air traffic controllers, nuclear power plants, and the US Navy aircraft carriers (Roberts, 1993). Consequently, high reliability organization theory has become a prominent model to explain complex high hazard organizations that perform with a high level of safety, reliability, and system integrity. High reliability organizations and safety culture share key attributes. Weick and Roberts (1993) suggest that organizations become reliable by creating a positive safety culture and reinforcing safety-related behaviors and attitudes. Research on high reliability organizations strengthened the case for safety as a facet of organizational culture.

1.2 TWENTY-FIRST-CENTURY LEADERSHIP AND CULTURE

Leadership is acknowledged as a primary influence on organizational culture (Schein, 2004). It has been my experience that impactful leaders have an innate talent for developing an inspiring picture of a future environment that is based on their personal core beliefs and their vision for an impactful outcome. They are able to paint a vision that motivates people to action. Leaders connect with people and inspire them to rally around an effort. Leadership requires a strong sense of self. Leaders help people understand how they can contribute to a greater good and in the process develop themselves. Are these leadership characteristics sufficient to influence organizational change in the twenty-first century?

In recent years, traditional leadership relationships have been replaced with non-traditional leadership processes and complex interactions. The connectedness, support, and development of staff is more difficult. Organizations are more fluid. Extended enterprises have emerged that are independent and networked. Cultural context has become important as the lines of nationality have become blurred and culturally acceptable norms are shifting universally.

Changing markets, flatter, more fluid organizations, global partnerships, and extended enterprises have created a plethora of unique roles for twenty-first-century leaders. Weaknesses in leadership tied to safety culture and high reliability attributes are important contributing factors in the analysis of catastrophic events. Retrospective case studies of disasters highlight that after a certain point within complex systems, mature safety processes and technology alone cannot assure safety; additionally, and more importantly, the influence of organizational and cultural factors associated with leadership must be considered. Rather than thinking of leadership narrowly in terms of the leader–follower perspective, we must consider the complex relationships and interactions that are necessary to influence positive outcomes in the twenty-first century and how they influence the organizational culture. Six disasters are described throughout the book:

- The explosion and release of toxic gas from a Union Carbide plant in Bhopal, India, that killed thousands of Indian citizens and injured hundreds of thousands in 1984
- The loss of seven crew members in the Challenger disaster when an O-ring failure caused the liquid hydrogen-oxygen tank to explode in 1986
- An explosion and fire at the BP Texas City Refinery that killed 15 workers and injured 200 in 2005
- An underground explosion at Massey Energy's Upper Big Branch Mine in West Virginia that killed 29 miners in 2010
- The release of radioactive contamination following the earthquake and tsunami at the Japanese Fukushima Daichii Nuclear Power Plant in 2011
- The explosion on the Deepwater Horizon rig that released approximately three million barrels of oil into the Gulf of Mexico, killed 11 workers, and injured 17 in 2012

Socio-technical factors influencing the six disasters are used throughout the book to support the case for authentic leaders that cultivate reliable and resilient organizations. Authentic leadership was born from society's need to trust government and corporate motives given the constant scandal and turmoil associated with recent catastrophic industrial disasters and other events with global impact (Northouse,

2013). Authentic leadership focuses on values and building trust in the organization through transparent relationships, which aligns stakeholders and ultimately influences the organizational culture (Avolio et al., 2004).

1.3 A NEW GENERATION OF CHALLENGES

Although valid and necessary, the traditional elements of safety programs are insufficient to mitigate catastrophic incidents and achieve the desired level of performance in all aspects of business. These elements need to be understood in terms of organizational risk and driven by the culture of the organization. Studies of world-class organizations have shown that there is a path to achieve leadership excellence in safety-related disciplines and ultimately improve performance across all business elements. Figure 1.1 describes the historical stages of development that have progressed our worldview of safety to one of collaboration that is focused on long-term organizational performance and continuous improvement. Authentic leadership is the silent factor that underpins success.

The following chapters apply the concepts of safety culture, high reliability, and authentic leadership to work teams operating in high-risk environments. They propose behavior-based risk management as a method to complement traditional activity-based approaches to reducing risk. The book's format will give the reader a better understanding of the characteristics of the leader–work team relationship that could prevent severe failures and offers practical tools to apply in the workplace. Although the focus is on environment, health, safety, and security, the concepts can be applied to any high-risk work environment.

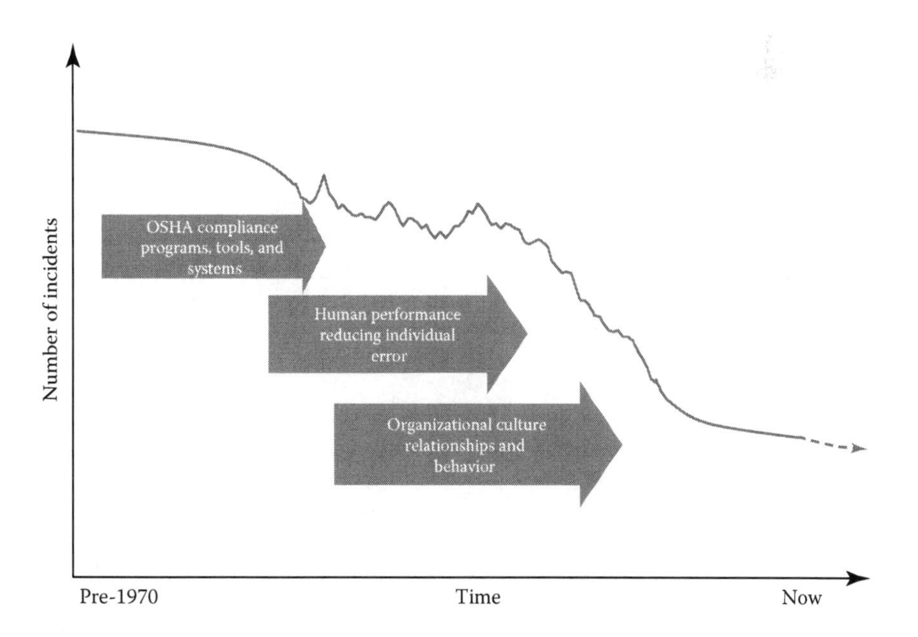

FIGURE 1.1 Historical stages to a strong organizational safety culture.

The book is organized into two sections:

- Section 1 provides a review of the literature and a case study.
- Section 2 is a practical toolkit for creating the capacity for organizational resilience and assessing, monitoring, and improving the organization's safety culture in the context of leadership.

Chapters 1 and 2 provide the historical background and a review of the literature. The literature review introduces a meta-model for leadership and work team engagement that considers leadership theory, accident causation theory, and safety culture research. Chapter 3 describes a case study that applies the concepts of safety culture, high reliability, and authentic leadership to the leadership of high-risk work teams. Chapter 4 describes organizational resilience and provides techniques to increase organizational capability through risk management and other organizational practices. It also describes a model to predict high-risk workgroups with a higher probability for future environment, safety, and health incidents. The model integrates aspects of employee engagement, safety culture, accident history, and exposure to high-risk activities to provide a unique lens to identify and manage safety risk across the organization. Finally, Chapter 5 outlines practical steps, including pitfalls and good practices, to assessing and improving safety culture and leadership within an organization.

Most importantly, the book provides a unique authentic leadership perspective associated with culture and organizational resilience that is based on a genuine leadership commitment to safe operations throughout the organization. Without such a commitment, the culture and resilient characteristics will not become ingrained in the values and beliefs of the organization and will be seen as nothing but shallow gestures by the workforce. Success is measured by a reduction in risk and increased organizational resilience to accidents.

2 Review of the Literature

This chapter comprehensively examines relevant leadership theory, accident causation theory, and safety culture research. Within each area, I discuss the current state of theory, research outcomes, and gaps in the literature. The review connects theory and research studies in these three areas to leadership behavior, work team engagement, and safety culture outcomes. The review also establishes a basis to better understand the nature of social relationships necessary to detect the precursors of catastrophic failures.

Using the traits and processes within each of these areas, I have proposed a metamodel (Figure 2.1) that describes the properties of an organization that routinely deals with hazardous technologies and does so with success and few failures. These essential elements to success are introduced in this chapter and applied throughout the book.

2.1 LEADERSHIP AS THE PRIMARY INFLUENCE ON CULTURE

Leadership is commonly accepted as the primary influence on organizational culture and safety culture (Schein, 2004). Retrospective case studies of disasters such as those that occurred at Bhopal (Shrivastava, 1987), on the Challenger (Vaughn, 1996), at the Texas City Oil Refinery (Hopkins, 2010), and on the Deepwater Horizon drilling rig (DHSG, 2011) have emphasized leadership's influence on organizational culture. Although much of the safety leadership literature aligns with the theories of leader–member exchange theory and transformational leadership, these theories and associated research are necessary but not sufficient to account for the complex and multidimensional nature of leadership.

2.1.1 Leader–Member Exchange Theory and Reciprocation

The concept of reciprocation associated with the leader–member exchange theory plays a role in safety leadership by enabling a positive work climate and positive safety outcomes. Basford et al. (2012) pointed out that the process of behavioral reciprocity asserts that people tend to respond to others' behaviors in kind. Immediate managers who feel supported by senior managers will reciprocate with supportive behaviors, and senior managers will respond favorably by showing support in return for displays of immediate manager support. Immediate supervisors that feel supported by their senior management may be more likely to forward this support to followers (Basford et al., 2012).

Hofmann and Morgeson (1999) identified that leader–member exchange and perceived organizational support influence accidents. Their research suggests that individual commitment to and engagement in safety is more likely when employees perceive organizational support and have a quality relationship with their supervisor.

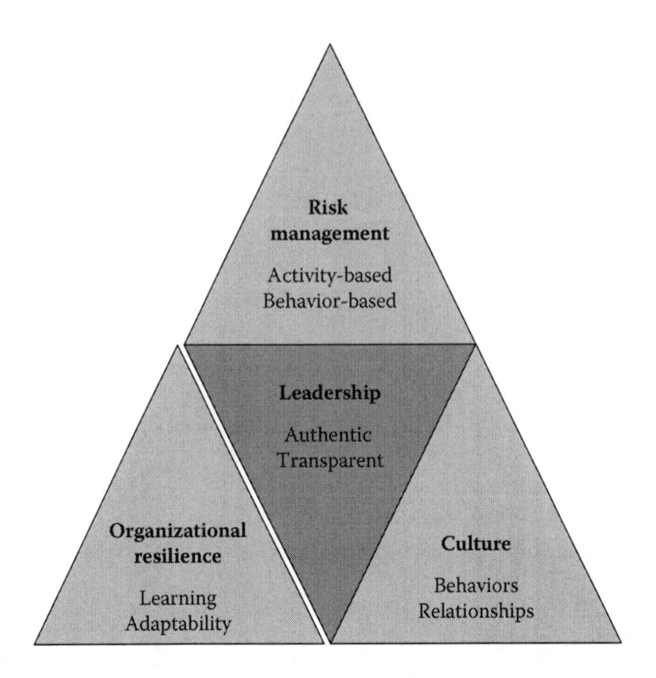

FIGURE 2.1 Meta-model for leadership in high-risk environments.

Michael et al. (2006) studied production employees at five wood manufacturing plants. They found that organizations that foster positive social exchange between employees and direct managers have less safety-related events. These studies provide important insight into transactional exchanges between managers, supervisors, and workers in specific contexts, but do not account for a leader's influence on cultural factors such as the values and beliefs of members.

2.1.2 Transformational Leadership Behaviors

Transformational leadership behaviors are credited with aligning values, increasing followers' trust, satisfaction, and citizenship, and predicting occupational injuries (Barling et al., 2002; Bass, 2008). Bass (2008) has been the primary influence on transformational leadership research. Bass believes that transformational leadership motivates followers to transcend their own self-interest for the good of the organization. The leader increases intrinsic motivation by increasing the perception that task objectives are consistent with their followers' authentic interests and values and moves followers to go beyond expectations by addressing their needs and motives (Bass, 2008). In essence, the follower's personal values are transformed to meet organizational goals.

The Multifactor Leadership Questionnaire (MLQ) has been the primary tool used to study transformational leadership (Jung & Avolio, 2000; Bass & Avolio, 1990). In a controlled study, Jung and Avolio (2000) found that transformational leadership behaviors had both a direct and indirect effect on the quality and quantity of follower performance. The effects were mediated by the follower's trust in the leader and value congruence between leader and follower (Jung & Avolio, 2000). In another study, Barling et al. (2002) used a modified version of the MLQ to assess follower perception of direct supervisors' transformational leadership behaviors in the restaurant industry. They found that safety-specific transformational leadership behaviors predicted occupational injuries (Barling et al., 2002). These concepts are important as the alignment of values and trust in the leader may enable the followers to persist and overcome significant obstacles, especially in high-risk environments.

2.1.3 Authentic Leadership

At the same time that the concept of safety culture was gaining momentum in the 1980s, leadership theories emerged that are based on a positive moral perspective. These theories build on Burns' (1978) concept of the moral leader that considers the essential desires and needs, hopes, and values of the followers. Authentic leadership is at the core of positive leadership. Avolio et al. (2004) describe authentic leadership as a root construct that can incorporate aspects of other positive leadership theories such as transformational leadership.

Authentic leadership enhances the transformational model by emphasizing authentic behaviors, role modeling, and value-based leadership that create an inclusive, ethical, and caring climate. Authentic leadership does not consider behavioral style but rather focuses on values and convictions, building trust among followers which aligns stakeholders and ultimately influences the organizational culture (Avolio et al., 2004). Gardner et al. (2005) integrated authentic leader perspectives into a self-based model of authentic leader and follower development. Three of the model's distinguishing features are balanced processing, relational transparency, and an internal moral compass. These features are important to establishing work team engagement and reducing system failure.

2.1.4 Authentic Leadership, Safety Culture, and High Reliability Theory

Authentic leadership addresses the leader's role in developing an organization's capacity to anticipate problems by engaging with staff to detect weak signals of failure through the notion of balanced processing (Gardner et al., 2005). Balanced processing refers to the handling of information and decision making that considers all opinions and encourages diverse viewpoints (Gardner et al., 2005). Similar to the characteristics of high reliability and safety culture, leaders request information, explore alternative solutions, challenge prevailing views, and welcome criticism (Eid et al., 2012).

The important underlying difference not explicitly captured in the leadership characteristics of high reliability and safety culture is that authentic leaders are driven by a moral compass and not influenced by external pressures (Walumbwa et al., 2008). Leaders that act ethically and morally establish trust within the organization and with the public. These leaders are motivated by their values and beliefs that reflect a higher mission. The following section reflects on leaders whose values are not aligned with the moral responsibility for the stewardship of worker safety and the environment.

2.1.5 Moral Reasoning: Authentic Leaders Have an Internalized Moral Perspective

In high-risk organizations, leadership driven by insincere motives has had disastrous consequences. A lack of passion for the environment and safety and health by senior executives resulted in misguided corporate priorities and values. In each of these examples, the corporate executives placed profit above their stewardship of workers, the public, and the environment. As disasters unfolded, their words and actions betrayed their true motives. Ethical dilemmas in times of duress can reveal a leader's true values. In public forums, the corporate Chief Executive Officers (CEOs) for Massey Energy and BP appeared very proud of their safety record and publically espoused their values for safety, but their behaviors sent a very different message to workers and the public.

The explosions at the BP Texas Refinery and the Deepwater Horizon oil rig are well-known examples of how senior executives emphasized worker safety (i.e., slips, trips, and falls) over the risks associated with industrial processes involved in refining oil and deepwater drilling. In both the BP Texas City and the Deepwater Horizon disasters, meetings were held with operations personnel to congratulate them for their safety record minutes before the disasters occurred. Both of these disasters highlight misplaced values at the expense of worker lives and environmental stewardship. Tony Hayward succeeded CEO Lord John Browne after the Texas City event. Browne was known for his hard-nosed cost-cutting tactics. When Browne left, many of the company's directors left as well (Bryant & Hunter, 2010). Many thought that Hayward's selection was a sign that BP was truly serious about improving safety standards. In one of his first interviews after being named BP's CEO, Hayward pledged to make safety a top priority and that he would "focus like a laser" on safety (Rayner, 2010).

Unfortunately, cost cutting continued at the expense of safety and set the stage for Deepwater Horizon. One month after the gulf disaster, Hayward dismissed the spill by stating "The Gulf of Mexico is a very big ocean. The amount of volume of oil and dispersant we are putting into it is tiny in relation to the total water volume" (Faulkner, 2010).

Likewise, when asked about Massey's safety record after the explosion that killed 29 miners, CEO Don Blankenship justified the loss of life by stating that Massey Energy employees worked in "difficult underground conditions" and described the 23 miner deaths in Massey Energy mines in the previous 10 years as "average" (Governor's Independent Investigation Panel, 2011, p. 93).

Blankenship maintained that safety had been his first priority since he joined Massey. He touted Massey as "an innovator of safety enhancements," a company that "has introduced many safety practices that have later been adopted throughout the mining industry in the United States and around the world" (Governor's Independent Investigation Panel, 2011, p. 94). Similar to BP's neglect of process risk, Massey Energy executives placed emphasis on their commitment to safety by requiring personal protective equipment such as reflective clothing, metatarsal boots, and seat belts, but neglected the fundamentals of safe mining such as suitable ventilation, operable equipment, and fire suppression (Governor's Independent

Investigation Panel, 2011). Multiple investigations of the Upper Big Branch disaster found that worn and broken equipment created a spark that ignited a build-up of coal dust and methane gas. Broken water sprayers could not prevent a minor flare-up from becoming a massive fire (Mine Safety and Health Administration, 2011; Governor's Independent Investigation Panel, 2011).

Focusing on accident statistics and protective equipment were superficial gestures to the employees that worked with unmaintained and malfunctioning equipment. Leadership can be described as a set of competencies, but it also requires a character and moral fiber that are unwavering in times of crisis. Unfortunately, leaders that are committed to moral values as well as those that are not influence the behavior of the entire organization.

2.1.5.1 Government Regulation: A Culture of Collusion?

As I discussed earlier in Chapter 1, US government regulation through the Occupational Health and Safety (OSH) Act has been responsible for significant improvements in workplace safety. Unfortunately, pandering with regulators and governmental agencies in order to put the corporation's self-interest above public health and welfare is another factor that has played a significant role in catastrophic events associated with highly regulated industries. Deepwater Horizon and Fukushima are examples that illustrate the consequences when boundaries are blurred between the functions of regulators and the contractors that they oversee.

In the case of Deepwater Horizon, the Minerals Management Services (MMS) was not only responsible for offshore leasing and resource management, it also collected and disbursed revenues from offshore leasing, conducted environmental reviews, reviewed plans and issued permits, conducted audits and inspections, and enforced safety and environmental regulations. Cross-purpose functions within the MMS created a conflict of interest that resulted in pressure from political and industry interests such as BP. The close relationship between BP and the regulators made the permit change process essentially a rubber stamp. It is likely that the approval of a permit to revise BP's drilling plan within 10 minutes of submission only helped to accelerate the inevitability of the explosion and environmental disaster (Roe & Schulman, 2011).

Similarly, in the case of the Fukushima accident, the regulation of nuclear power was entrusted to the same government bureaucracy that promoted it. The nuclear disaster was determined to be the result of complicity between the government, regulators, and operating contractor, and their collective lack of governance. The operators and regulators put their own interests in expansion and profits ahead of their responsibility to protect the safety of the public. Stakeholders involved in Fukushima were well aware of the best practices that had been advocated by international agencies, but they failed to force the industry to implement them and covered up small scale accidents (DIET, 2012).

To assure public safety and environmental protection, the separation of the regulatory function charged with promotion and the function charged with production requires reforms even beyond those already enacted since the Deepwater Horizon and Fukushima disasters. Fundamental reform is needed in both the structure of those in charge of regulatory oversight and their internal decision-making process to ensure their political autonomy, the use of technical expertise, and their full consideration of environmental protection concerns (Deepwater Horizon Study Group, 2011; DIET, 2012).

2.1.5.2 Conclusion

It is difficult to measure the institutional benefit of averting disaster, but the cost of major industrial disasters is a topic of scholarly research and a matter of public record. Authentic leadership is not about one decision to avert disaster, it is a way of doing business throughout the enterprise that cultivates and builds the character of a corporation over time. Authentic safety leadership for high-risk corporations is complex and must be aligned at all levels within the organization and reflected in interactions with all stakeholders.

To be successful in the twenty-first century, CEOs in complex high-risk industries must understand that profitability is tied to the protection and preservation of human life and the environment. Corporations such as Massey Energy and BP believed that loss of life and environmental insult were part of the cost of doing business. Leadership in high-risk industries must make decisions that reflect care and compassion for human life without letting external influences tied to profit and legal exposure get in the way.

2.1.6 Authentic Leadership, Psychological Capital, and Work Team Engagement

Kahn (1990) defined engagement as "the harnessing of organizational member's selves to their work roles; in engagement, people employ and express themselves physically, cognitively and emotionally during role performances" (p. 694). Kahn (1990) argued that psychological safety enables personal engagement at work by establishing trust and respect. Edmondson (2012) found that people working together in teams tend to have the same set of influences and perceptions, and introduced the term *psychological safety* as a group-level construct. In psychologically safe environments, teams are comfortable and willing to offer ideas, question anomalies, and raise concerns (Edmondson, 2012). Similar to the characteristics of psychological safety, authentic leadership supports work team engagement by developing positive psychological capital through sharing information in an open and transparent manner (Eid et al., 2012).

Psychological capital is a concept that played an important role in the conceptualization of authentic leadership (Luthans et al., 2015). Psychological capital refers to an individual state of development that inspires individuals with confidence to succeed at challenging tasks, work toward challenging goals, and rebound from adversity (Luthans et al., 2015). Luthans and Avolio (2003) identified four positive psychological states: confidence, optimism, hope, and resiliency. Luthans et al. (2004) suggest that psychological capital can play a role in developing teams and organizations. These claims are supported by research done by Walumbwa et al. (2010). Walumbwa et al. (2010) found that the more leaders were seen as authentic, the more employees identified with them, felt empowered, engaged in work roles, and demonstrated organizational citizen behaviors. Relational transparency and the development of positive psychological capital appears to be closely related to positive organizational outcomes. Eid et al. (2012) reflect that developing the positive psychological capital associated with authentic leadership becomes important in complex high-risk industries that must adapt to unpredictable events.

2.1.7 Leadership and Culture

The theories of leader–member exchange and transformational leadership as enacted through authentic leadership are important factors that are thought to contribute to safe performance and work team engagement. Authentic leadership actions demonstrate a genuine passion for the well-being of the organization, align values with their followers, and promote the reciprocation of behaviors (Walumbwa et al., 2010). Unfortunately, although empirical research has demonstrated that leadership can predict a variety of outcomes such as performance and employee attitude, it usually accounts for less than 10% of the variance in these outcomes (Eberly et al., 2013). These studies primarily focus on the relationship between leader and follower and fail to consider the interactive relationship between leader, follower, followership, and the organization. Eberly et al. (2013) suggest that a more integrated model is needed to fully explain the mechanisms influencing leadership outcomes. They suggest that research has underspecified context and that a multidimensional integrated view is needed to account for greater variance in outcomes. Their model considers a diffused process that is independent of any formal role or hierarchical structure. They describe leadership as a series of dynamic event cycles between multiple loci of leadership. The model considers the dynamic interplay between leaders, followers, dyads, and collectives and how they influence organizational outcomes. Until recently, literature has been silent on the dynamic interplay of leadership (Eberly et al., 2013). This is an important consideration to fully understand the impact of leadership in establishing a resilient culture that actively detects the precursors of catastrophic accidents.

2.2 CATASTROPHIC EVENTS IN HIGH-RISK ENVIRONMENTS

Catastrophic events are low-probability, high-consequence occurrences that have a high inherent risk (Kleindorfer et al., 2012). In recent history, catastrophic events have devastated ecosystems, nations, industries, and communities. There are two schools of thought about preventing catastrophic accidents. The first, normal accident theory, contends that in highly coupled complex high-risk industries, accidents are inevitable and unpredictable (Perrow, 1999). The second, high reliability theory, focuses on organizational resilience and reliability to avoid normal accidents. High reliability depends on the organization's ability to detect and remedy small deviations from expected operations before they combine and contribute to a catastrophic event, and is a significant factor in establishing a strong safety culture and highly engaged work team.

2.2.1 Man-Made Disasters

Turner's (1978) model for man-made disasters is based on a failure of foresight to detect a drift from organizational norms. Turner (1978) believed that prior to disaster there is a long incubation period where the potential for disaster builds. The model was groundbreaking as it defined disasters in sociological terms as a collapse of existing cultural beliefs about hazards. The theory posits that system vulnerability arises from unintended and complex interactions between seemingly normal organizational functions. It also asserts that the best safety processes and systems could be undermined by organizational behaviors that suppress or dismiss information.

Karl Weick (1987) opined that failures occur mainly through psychological factors and used the term *requisite variety* to explain why humans do not possess the capacity to anticipate problems generated by complex systems. Weick (1987) argued that the variety of the complex system exceeds the capability of the human brain to sense and anticipate problems. When the brain's capability is overloaded, important information is missed which can exacerbate a problem. High reliability research expanded Turner's model through the contention that careful, mindful organizational practices can make up for inevitable limitations to the rationality of individual members.

Weick and Sutcliffe (2001) proposed several high reliability concepts to enhance the organization's ability to sense and anticipate problems. Although many of the studies have comprehensively described the theoretical basis of high reliability concepts in terms of characteristics, little of the research has tied cultural characteristics to organizational performance (Weick & Sutcliffe, 2001, 2007).

2.2.2 High Reliability Theory

Mindfulness is a foundational basis of high reliability. Mindfulness is characterized by a questioning attitude that accepts divergent viewpoints and an awareness of operations that are also associated with a strong safety culture. Mindfulness is a state of constant awareness within high hazard environments, looking for subtle indications of failure (Weick & Roberts, 1993). Organizational members are able to improvise and quickly develop new ways to respond to unexpected events. High reliability organizations might experience numerous failures, but their resilience and swift problem-solving helps to prevent catastrophes. Mindfulness is the backbone of two high reliability concepts that leaders use to reduce the likelihood of failure.

The first concept is coined *preoccupation with failure* and refers to the constant preoccupation with potential errors and failures (Weick & Sutcliffe, 2007). Organizations practicing preoccupation with failure use incidents and near misses as indicators of a system's health and reliability. Organizations that are preoccupied with failure systematically collect and analyze warning signals. Leaders effectively anticipate problems by engaging with frontline staff in order to obtain the bigger picture of operations and being attentive to even seemingly minor or trivial signals that may indicate potential problem areas within the organization (Weick & Sutcliffe, 2007). Leaders worry chronically and help employees worry chronically about errors. They assume each day is a bad day. Leaders pay close attention to operations but don't micromanage. Leaders make sure everyone values organizing to maintain situational awareness.

A second closely related reliability-enhancing concept is referred to as *reluctance to simplify*. Reluctance to simplify focuses on the organization's ability to avoid making any assumptions regarding the causes of failure by taking deliberate steps to create a more nuanced and complete picture (Weick & Sutcliffe, 2007). Organizations practicing the concept of reluctance to simplify assume that failures are systemic, rather than localized, and could potentially lead to a broader causal chain of events with potentially catastrophic consequences (Weick et al., 1999). Leaders understand that organizations must filter information and doing so may force them to ignore key sources of problems. High reliability leaders help employees restrain temptations to simplify by establishing checks and balances, adversarial reviews, and multiple perspectives.

Weick and Sutcliffe's (2007) concepts of preoccupation with failure and reluctance to simplify overlap with Roberts and Bea's (2001a) description of high reliability organizations' inclination to aggressively seek to know what they do not know. These behaviors are also reflected in the nuclear power industry's description of safety culture behaviors that ensure that problems are thoroughly evaluated by the organization to understand the organizational and safety culture contributors (INPO, 2013). Similarly, the safety culture attributes used by the Department of Energy (2011) describe this behavior as *operational awareness*. Operational awareness is defined as line managers listening to and acting on real-time information by maintaining close contact with those that are doing the work (DOE, 2011). Each of these descriptions capture the ability to collect, analyze, and synthesize information about the bigger picture of current operations in such a way that enables them to effectively contain and prevent potential future failures.

The concepts of operational awareness, preoccupation with failure, and reluctance to simplify are necessary as failure is not an option in most high-risk industries. High reliability organizations do not usually adopt a trial-and-error strategy because the political, economic, and institutional costs of errors are unlikely to be offset by the benefits (Boin & Schulman, 2008). High-risk industries rely on learning from events that have a low impact on the organization but also touch on organizational concerns such as near misses and minor accidents (Lampel et al., 2009). Hopkins (2010) argued that high reliability leaders use accidents that happen in other organizations as opportunities to check whether similar problems exist in their organization.

A limitation to studying retrospective events is that they force us to view history through the lens of hindsight, which makes it difficult to understand all of the variables that might have caused the error. High reliability organizations value and reward the reporting of near misses and errors because they are viewed as learning opportunities and a means of obtaining a realistic picture of operations. Further, near misses are thoroughly analyzed because they are seen as opportunities to improve operational processes.

2.2.3 The Influence of Leadership in Catastrophic Events

Most of the studies associated with accident causation are retrospective case studies of disaster. Within the context of leadership, these disasters were not causally linked to one bad decision, but instead were socially organized and systematically produced by bureaucratic social structures.

- The Bhopal disaster exemplifies management's failure to engage at all levels within the American parent corporation, its Indian subsidiary, and the local Indian government. Union Carbide moved out of Bhopal to give the Indian government self-sufficiency and local control (Browning, 1993), but instead of collaboration and cooperation the relationship was strained. The local government was aware of safety problems but was afraid to act for fear it would lose a large employer (Broughton, 2005). Information and concerns were stifled at all levels within the corporate and subsidiary organizations.
- The Challenger disaster may have been averted had the organizational culture valued open reporting and the thorough analysis of issues raised by staff. After the incident, senior NASA decision makers told the Rogers Commission that they had no idea of the controversy between the contractor and Space Center senior management the night before the launch (Vaughn, 1996). Clearly, there was no integrated process in place to evaluate differing opinions across functions within the organization, but rather a tendency to resolve problems in-house at the lowest level (Hauptman & Iwaki, 1990).
- Failures in management, budgetary priorities, and corporate values were noted as primary contributors to the BP Texas Oil Refinery disaster (Hopkins, 2010). "None of the management accountability failings identified by the team caused the disaster. Rather, the culture present at Texas City Refinery was the single most direct causal connection" (Hopkins, 2010, p. 140).
- In the case of Deepwater Horizon, the crew's good injury record and success associated with their previous deepwater drilling assignment which set world records likely gave leadership an inflated confidence in the engineered controls and led to complacency (DHSG, 2011).

These examples highlight the importance of leadership's role in establishing an organizational culture that could prevent catastrophic events within high-risk environments. The following discussion applies the Bhopal, Challenger, and Deepwater Horizon events to emphasize weaknesses in problem anticipation, mutual engagement, just culture, and organizational learning that contributed to disaster. Figure 2.2 illustrates the relationship between leadership and these four essential elements of safety culture.

2.2.3.1 Problem Anticipation

The Deepwater Horizon disaster had precursors that went unnoticed or were ignored. In complex systems, a single safeguard can fail with no effect because of redundant safeguards. A build-up of these preexisting conditions makes the system vulnerable to catastrophic failure.

The parent company of Deepwater Horizon, BP, as well as its contractors grossly neglected warning signs that indicated problems. Errors and misjudgments by three companies—BP, Halliburton, and Transocean—contributed to the disaster. Warning signals were ignored by senior management who repeatedly approved inexpensive, less stable well designs, as well as frontline management who dismissed clear warning signs and jeopardized the crew's safety to meet deadlines (DHSG, 2011; Safina, 2011).

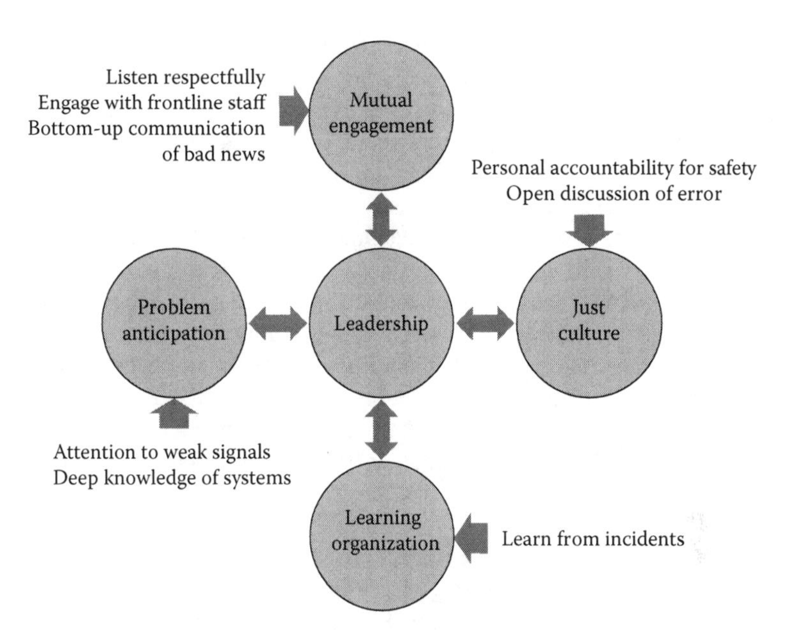

FIGURE 2.2 Leadership and essential safety culture elements.

One potential barrier to problem anticipation is the nature of safety that can create an erroneous belief of infallibility. Weick and Sutcliffe (2001) posit that "Safety is elusive because it is a dynamic non-event—what produces the stable outcome is constant change rather than continuous repetition" (pp. 30–31). The problem is that when a system is operating safely and reliably there are constant outcomes that do not attract attention. Safety is constantly moving and changing though nothing seems to happen. Weick and Sutcliffe's (2001) supposition matches my experience as a reactor operator in the late 1970s. My job was to look for signs of failure within the reactor operating systems by sitting at a console and monitoring several panels of stable indicators for an entire shift. It was easy to be lulled into a sense of status quo when in fact the system was in a mode of continuous adjustment.

The Deepwater Horizon crew regarded the absence of change as safe and reliable. Hindsight suggests that the reality was much different. It is likely that both the operators and leadership on the Deepwater Horizon rig may have been one decision away from disaster on many previous occasions. They passively accepted safe outcomes instead of being engaged in detecting anomalies.

2.2.3.2 Mutual Engagement

Mutual engagement between leadership and followership is an underlying characteristic of both safety culture and high reliability. Engaged leadership attributes associated with a strong safety culture are identified with teamwork, by actively seeking the opinions and concerns of workers at all levels of the organization, respecting differing opinions, and not tolerating conditions that could reduce operating safety margins (INPO, 2013; DOE, 2011). Similarly, drawing on high reliability organization (HRO) research findings, Hopkins (2010) argued that leadership must encourage "bottom-up" communications and Roberts and Bea (2001a) identified that organizations can enhance their reliability by communicating the big picture to everyone. The linkage between positive psychological attributes such as engagement and safety outcomes is important because it indicates that employees are applying their whole selves to their working role and are cognitively and emotionally engaged in detecting anomalies that may be precursors to failure.

For the companies associated with Deepwater Horizon, workforce engagement and open two-way communication was not the norm. The report issued by the National Commission on the BP Deepwater Horizon Oil Spill and Offshore Drilling found that excessive compartmentalization of information between BP, Transocean, and Halliburton hampered communication. Both BP and its contractors did not share important information with each other. BP was even closed off with members of its own team. Poor communication resulted in decisions being made without a full appreciation for the context of the situation or recognition that the decisions were critical (National Commission on the BP Deepwater Horizon Oil Spill and Offshore Drilling, 2011).

For example, at the time of the explosion, BP was in the process of temporarily abandoning the well. A long string case design was chosen as part of the procedure to seal the well. Centralizers on the casing keep it in the center of the wellbore so that cement can be poured around the casing. If it is not properly centered, the wellbore will not fully seal and could initiate a blowout. Days before the event, BP well engineers deliberated on the number of centralizers to be used on the casing. Although BP's original plan called for 21 centralizers, the casing available on the rig had only 6 centralizers. A concerned Halliburton engineer recalculated and validated the required 21. The casing was ultimately installed with 6 centralizers by proclamation of the BP project manager (DHSG, 2011).

Leadership is critical to establishing mutual engagement through the modeling of behavior, actively seeking the opinions of workers, and respecting differing opinions. These behaviors if genuinely and authentically expressed will develop the positive psychological resource capacity of staff. In addition to the barriers previously discussed, what organizational factors might discourage mutual engagement?

A significant barrier to achieving the mutual engagement of leaders and followers is organizational bureaucracy. Hudson (1999) characterizes a bureaucratic culture as one where information is ignored, messengers are tolerated (rather than welcomed or shot), new ideas are seen as problems, and responsibilities are compartmentalized. In bureaucratic organizations, thinking may stop at the department's boundary because what is beyond it is "not my concern," even though it may be of great concern to the mission. The ability to process information is often closely tied to the ability to act on the information and responsibility becomes compartmentalized (Hudson, 1999).

In a bureaucratic organization when opinions and advice are not solicited or welcomed, the lack of balanced decision making can result in a phenomenon called "normalization of deviance." The term was first coined when referring to NASA's culture that contributed to the 1986 Challenger accident (Dekker, 2011). Normalization of deviance refers to a gradual process through which unacceptable practices or standards become acceptable. As the deviant behavior is repeated without catastrophic results, it becomes the social norm for the organization. Individuals who challenge the norm from within the organization or outside it are considered nuisances or even threats (Dekker, 2011).

Both Union Carbide and BP and their contractors operated under a normalization of deviance. For Bhopal, it gave them permission to continue to cut manpower and allow safety systems to fail. For Deepwater Horizon, it allowed impromptu design changes to save time and money. Both organizations had significant compartmentalization of functions that stifled differences of opinion and resulted in a lack of integrated knowledge about critical processes.

2.2.3.3 Just Culture

Without the prerequisites of open communications and personal accountability for safety, a just culture is not possible. In the case of the Bhopal incident, it appears that information and concerns were stifled at all levels within the corporate and subsidiary organizations. Leadership's action and inaction sent a loud message that the safety of the workers or the community was not valued. At least six serious incidents occurred between 1981 and the disaster in 1984. During this timeframe one worker died, several workers suffered extensive chemical burns, and dozens were hospitalized after exposure to gas. Management took no action to analyze the situation to reduce the risk of failure (Gupta, 2002). According to accounts, workers at the plant tried to voice their concerns. In 1982, labor unions complained to the ministry of labor about plant conditions and nothing was done. Incredibly, prior to the disaster, 70% of the plant's employees were fined for refusing to deviate from the proper safety regulations under pressure from management (IDEESE, 2009). Leadership of the Bhopal chemical plant appeared to have a lack of regard for life and their community. They had no processes or systems to report concerns and took strong steps to discourage reporting.

Management at the Bhopal plant had political as well as schedule and profit-driven motives to suppress voices that got in the way. In the case of Bhopal, there was strong political pressure to keep the jobs within the community despite the lack of funding to maintain critical safety systems and protect workers from chemical exposure.

There are also more subtle barriers to the development of a just culture in organizations than those demonstrated at Bhopal. In many organizations, individuals tend to resolve problems within their workgroup rather than report them, which limits open communication and learning from mistakes. For example, Provera, Montefusco, and Canato (2008) found that a just culture was suppressed in an intensive care unit that did not openly discuss errors within the organization but rather addressed errors only within individual teams.

Most are familiar with the Challenger accident and the O-ring argument between NASA management and contractor engineers prior to the decision to launch. What many may not know are the more subtle communications practices within the agency that contributed to the decision. In the case of the Challenger incident, contractor safety engineers continued to recommend launching on previous missions despite knowledge of worsening erosion and heat damage to the O-ring. They viewed the risk as acceptable so never reported their concerns until the night before the Challenger launch (NASA Safety Center, 2013).

Suppressing information and resolving problems at the lowest level in the organization stifles the organization's ability to learn from failure. In this respect, just culture is very closely tied to organizational learning.

2.2.3.4 Organizational Learning

Organizational learning is the final characteristic that I believe supports a strong safety culture and high reliability organizations. In a strong safety culture, the attributes of organizational learning are associated with critical assessment, learning from incidents, and assuring knowledge transfer and the retention of experienced staff. Leadership sets the tone for organizational learning. Executives and senior managers of learning organizations seek outside perspectives and support candid thorough assessments of their operations. Leaders openly communicate the results of monitoring and assessment throughout the organization. Knowledge transfer and knowledge retention strategies are applied to capture the knowledge and skill of experienced individuals (INPO, 2013).

Drawing on HRO research findings, Hopkins (2010) argued that leaders in learning organizations use accidents that happen in other organizations as opportunities to check whether similar problems exist in their organization. Leaders also proactively commission audits to diagnose any weaknesses in the organization's defenses and question audit findings that only deliver good news. HROs value and reward the reporting of near misses and errors because they are viewed as learning opportunities and a means of obtaining a realistic picture of operations. Further, near misses are thoroughly analyzed because they are seen as opportunities to improve operational processes.

Similarly, authentic leaders encourage organizational learning through words and actions that allow the organization and its members to grow. Through positive modeling of both transparency (openness, self-disclosure, and trust) and balanced processing (unbiased interpretation of information), leaders develop authentic followers (Gardner et al., 2005). Modeling occurs using behaviors that reflect high moral standards and promote innovative problem-solving (Luthans & Avolio, 2003). In time, trustworthiness, honesty, integrity, and accountability become shared values of the organization (Gardner et al., 2005).

Clearly the organizations associated with the Deepwater Horizon explosion and the Bhopal chemical disaster lacked the behaviors and beliefs necessary to promote organizational learning.

For BP and its contractors, decision making at all levels reflected an unwillingness to learn from errors or investigate accidents. As I previously pointed out, management dismissed clear warning signals that could have been used as opportunities to learn. The repeated approval of changes that contributed to a less stable well design such as the decision to reduce the number of centralizers reflected an environment where collaboration and the sharing of previous experience was not welcome. Misunderstanding indications during a critical negative pressure test of the well seal exemplifies a limited understanding of the processes. Stovepiped information meant that no one had the big picture of what was happening with the well.

Similar to Deepwater Horizon, the Bhopal plant was focused on cutting costs, which allowed for the degradation of equipment, maintenance, and process knowledge. The capability of the workforce was severely degraded in 1984. Management viewed workers as expendable and did not value their knowledge. As skilled workers left they were replaced with less-educated workers. The methyl isocyanate plant staff was reduced from twelve to six (Cassels, 1993). The workers at the plant that night did not have the process knowledge to understand the consequences of their actions and were not aware of the severity of the event even after it occurred.

Sharing experience across organizational boundaries, investigating near misses and accidents, and assuring a skilled and experienced staff that have integrated system knowledge were not common practices. Missed opportunities to learn and share information and develop workers had disastrous consequences for the workers, their families, the surrounding community, and the ecosystem.

The phenomenon of hindsight failure portrays the past as more linear and foreseeable because we can start with the outcome and work back to the "cause." This tends to oversimplify the cause (Dekker, 2006). For example, in 1986, the Challenger exploded when an O-ring failed to seal at low temperature. There were warning signs of a problem with the O-rings.

Without the benefit of hindsight, preventing the Challenger disaster would have required evaluating every warning sign which appeared as the potential O-ring problem. Additionally, initiating events can also be overlooked through cognitive bias, that is, we don't see what we are not looking for (Simons & Chabris, 1999). Roe and Schulman (2011) commented that the complexity of technical systems challenges workers not only to find useful information in a mountain of data, but also to find needed information that they do not recognize.

An organization's willingness to change as a result of learning from an event is directly related to how the organizations perceive the impact to their operations (Lampel et al., 2009). Attention is differentially allocated depending on the priorities and issues that decision makers see as being central to their strategy. In the case of Bhopal and Deepwater Horizon, it was clear that the priorities and issues that the corporate leaders deemed essential were not related to safety.

2.3 ORGANIZATIONAL CULTURE, SAFETY CULTURE, AND SAFETY CLIMATE

The concepts of safety culture and safety climate are closely aligned with organizational culture. Most of the literature supports the notion that safety culture is not a separate function but rather an integrated part of organizational culture. In reference to the nuclear industry, Apostolakis and Wu (1995) suggested that the strong tie between safety culture and organizational culture is through strong common dependencies in work processes and organizational factors. Attributes of safety culture and dimensions of safety climate have a strong relationship with leadership and safety outcomes, which are important factors that contribute to organizational reliability and the prevention of failure in high-risk industries.

The terms safety culture and safety climate are interrelated. Guldenmund (2000) provided a comprehensive summary review of the range of available definitions. In comparing safety culture with safety climate, he suggested that safety climate refers to the prevailing attitude toward safety within an organization whereas safety culture concerns the broader underlying beliefs and convictions of those attitudes. Research often confuses the two. A difficulty associated with operationalizing safety culture in quantitative research is that it becomes conflated with safety climate. Regardless of their tight connection and interchangeable references in the literature, the two terms have developed into research areas with different epistemological and methodological approaches.

The majority of the safety culture models focus on how people think, that is, their attitudes, beliefs, and perceptions. Studies in safety culture use a functionalist approach that assumes culture exists as a discrete entity: organizational culture is a component of the organization, so the culture of the organization has a discrete purpose and function (Janićijević, 2011). A central assumption to the functionalist approach is that culture can be managed and changed to influence outcomes such as safety and reliability. This viewpoint is accepted by most psychologists, management consultants, and organizational theorists, and predominates the field of safety culture research to the present day (Guldenmund, 2010).

On the other hand, the interpretive approach is based on finding "meaning" as the main point of cultural analysis (Haukelid, 2008). Interpretive scholars argue that culture is created and recreated through members' interaction and negotiation over meaning (Pidgeon & O'Leary, 2000). In the interpretive approach, the researcher establishes meaning through the research based on input from the organization and becomes an instrument of the research. Methods include observation, analysis of interview transcripts, and a study of accident investigation reports (Haukelid, 2008; Janićijević, 2011; Turner, 1978). Interpretive cultural research typically provides a more comprehensive insight into the complexity associated with organizational culture.

The functionalist approach and use of questionnaires is the preferred method to study safety culture (climate) and its relationship to behaviors and safety outcomes. The practical attraction to this approach is that measuring precursors of accidents versus traditional lagging indicators such as days away from work could be a powerful management tool for preventing injury or a catastrophic event. The drawback to this approach is that it is superficial and lacks the depth to understand the multidimensional nature of culture. Despite the differences in approach, research examining safety climate has provided the most insight into the correlation between aspects of safety climate and safety outcomes.

2.3.1 Correlation between Aspects of Safety Climate and Safety Outcomes

Coyle et al. (1995) found a relationship between safety-climate factor scores and lost time accident rates. Lee's (1996) results showed a strong correlation between contentment with job, satisfaction with work relationships, and low accident rates. A study by the Health and Safety Executive (HSE, 2003) found statistically significant relationships between incidents such as self-reported injuries and dangerous occurrences. Johnson's (2007) study in a heavy manufacturing organization found that safety climate responses were an effective predictor of safety-related outcomes of safe behavior and accident experience. Bjerkan (2010) found a significant correlation between perceived work safety climate/perception of the physical and psychosocial work environment and health status/accidents. Christian et al. (2009) examined antecedents of safety behaviors and identified that group safety climate had the strongest association with accidents and injuries. Neal and Griffin (2006) examined perceptions of safety climate, motivation, and behavior and linked them to prior and subsequent levels of accidents over a five-year period. They found that average levels of safety climate within groups at one point in time predicted subsequent changes in safety motivation. In addition, improvements in the average level of safety behavior within groups were associated with a subsequent reduction in accidents at the group level.

These research studies relying solely on questionnaires demonstrate a relationship between safety-climate factors and safety outcomes, but they provide an incomplete understanding of culture. Regardless of the criticism, the studies are important because they provide insight into how the organization might impact outcomes by influencing attitudes and ultimately values and beliefs. Such influence primarily resides in the realm of leadership.

2.3.2 More on Safety Culture and Just Culture

Section 2.2.3.3 described just culture as an essential element of a strong safety culture. The elements of a just culture are also connected to high reliability concepts that encourage teams and individuals to learn from error. The concept of a just culture begins to shift the focus from assigning blame and punishment to the belief that human fallibility and at-risk behaviors are normal and predictable. To begin the shift in thinking, the organization must distinguish between unacceptable or blameworthy behavior and behavior that is reasonable given the circumstances (Reason, 2000). In simplest terms, a just culture balances learning from incidents with accountability for consequences.

The safety culture attributes identified with a just culture address reward and discipline. Leaders reward individuals that raise concerns and consider the repercussions on the organization prior to taking disciplinary action. Individuals promptly report degraded conditions and near misses with the expectation that unintended errors will be acknowledged positively. The aviation and healthcare industries have been actively promoting improvements in the quality and reporting of safety concerns in the context of just culture. The aviation industry established the Global Aviation Information Network (GAIN) to facilitate the voluntary collection of safety information in the international community and has shared case studies on establishing a just culture. In one case study, the Danish government proposed a law to make non-punitive, confidential reporting possible. After the first year, 980 reports were received in comparison to 15 the previous year (GAIN, 2004).

Just culture begins with individuals taking action. In organizations where safety is not an essential part of the mission or reason for existence such as a chemical production plant, there can be extreme pressure to achieve non-safety goals which pressures employees to ignore the rules to meet production deadlines or save money. On the other hand, in industries with safety-critical operations, staff may be reluctant to report for fear of punishment. A just culture promotes a questioning attitude, which fosters a sense of personal accountability for safety. Leaders place a high value on safety and expect that staff will step back and voice their concerns before continuing with their work. Westphal (2009) has cited reductions in maintenance errors in the airline industry and increases in hand hygiene compliance in the healthcare industry associated with just culture prevention strategies. In these cases, the just culture encouraged members to take personal responsibility for the safety of those that they serve by openly reporting errors.

Leadership is critical to establishing a just culture. Leaders must take deliberate steps to create open communication and focus on rewards versus blame when failures occur. Authentic leadership addresses the intent of a just culture through relational transparency. Leaders walk the talk and live by their own ethical and moral standards (Eid et al., 2012).

Establishing a just culture requires more than processes and procedures that address reporting, rewards, and disciplinary action, and is closely tied to the characteristic of mutual engagement. Instituting a sustainable just culture requires that the organization first create an environment of trust and openness. For example, Frankel et al. (2006) discussed the importance of open communication in improving the reliability of healthcare organizations. They pointed out that the predominant mind-set in hospital operating rooms regards each staff member as an expert in their field and posited that such an expert-based system for handling issues does not foster an environment where staff can openly raise concerns regarding a patient's care (Frankel et al., 2006). They proposed that an environment where each operating room team member can voice concerns and feel comfortable voicing concerns when they arise is key to reducing error rates.

2.3.3 Linkage: Management Dimensions and Safety Outcomes

As was previously discussed, leadership is commonly accepted as the primary influence on organizational culture and safety culture. So, it is not surprising that research has also linked specific management dimensions to safety outcomes. Cheyne et al. (1998) identified management commitment to be key in their predictive model of safety behaviors. Gershon et al. (2000) found the dimension of senior management support to be especially significant with regard to both compliance with universal precautions and exposure to blood-borne pathogens. Sawacha et al. (1999) researched factors affecting safety performance on construction sites and found that top management's attitude toward safety was a significant factor in safety performance as measured by accident rates. In several intervention studies, Zohar (2002) and Zohar and Luria (2003) identified a significant increase in the use of protective equipment in departments where the direct manager interacts frequently with staff about safety.

In addition to research that has shown a positive relationship between leadership commitment and support to safety and follower safety outcomes, research has also examined the less tangible social aspects between leadership and followership. Spreitzer and Mishra (1999) identified that trust is engendered through open communications, giving employees more decision-making abilities, and the honest sharing of feelings/perceptions. Luria (2010) explored how trust acted as an antecedent for the promotion of safety by soldiers in operational units. Trust was found to be negatively related to injury rate and positively related to safety-climate strength. Safety-climate level fully mediated the trust–injury relationship. Nahrgang et al. (2011) conducted a meta-analysis that explored how workplace demands and resources influence workplace safety (accidents and injuries, adverse events, and unsafe behavior). The study found that burnout was negatively related to working safely but that engagement motivated employees and was positively related to working safely. Across industries, risks and hazards were the most consistent job demands and a supportive environment was the most consistent job resource in terms of explaining variance in burnout, engagement, and safety outcomes (Nahrgang et al., 2011).

2.4 SUMMARY

History has revealed that complex high hazard industries cannot prevent catastrophic failures through reliance on technology alone and must consider organizational culture. Leaders have a significant influence on culture and leadership failures have contributed to major industrial catastrophes. The theories of authentic leadership, high reliability, and safety culture provide characteristics of leadership that promote trust and engagement, as well as resilient, safe operations. Leadership characteristics are described in terms of social interactions that encourage diverse viewpoints and the sharing of information openly and transparently to detect weak signals of failure. Figure 2.3 depicts the relationship between authentic leadership, high reliability/ safety culture principles, trust, and organizational resilience.

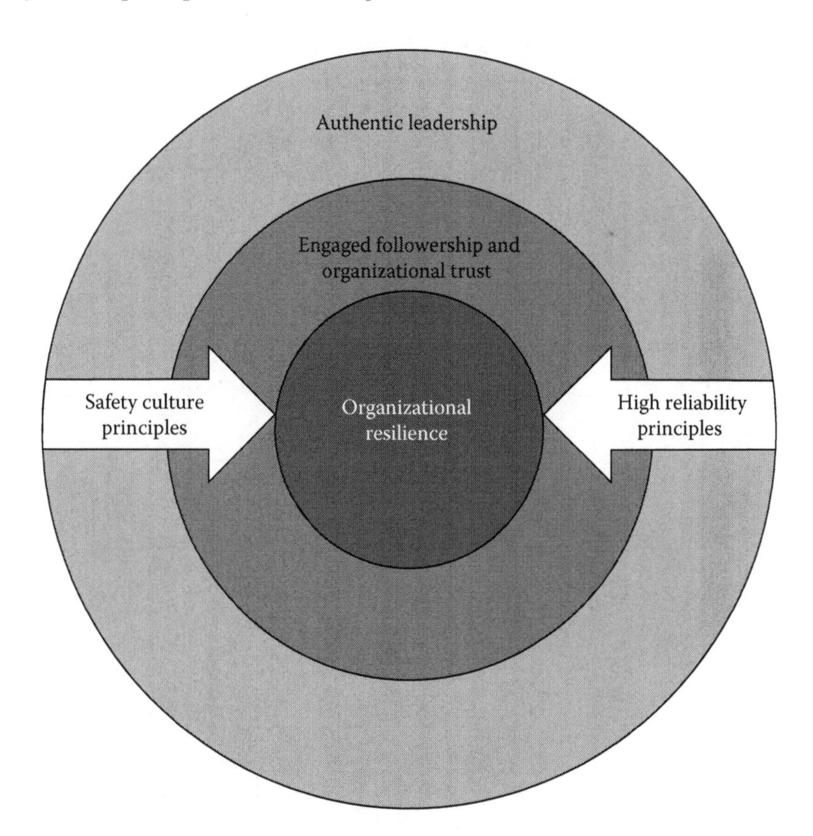

FIGURE 2.3 The relationship between authentic leadership, high reliability/safety culture principles, trust, and organizational resilience.

Thoughts on how to create the capacity for organizational resilience are examined in Chapter 4. The concept of organizational resilience is important in a high-risk setting as it can help detect the onset of a potentially devastating change in the working environment. Together, the concepts of authentic leadership, safety culture, and high reliability enhance the organization's capacity for resilience by detecting the onset of a potentially devastating change and facilitating recovery (Weick et al., 1999).

The majority of safety culture research has taken a functionalist approach that uses climate surveys to understand specific cultural dimensions (Janićijević, 2011). Studies have found a statistical correlation between single aspects of safety climate and safety outcomes (Bjerkan, 2010; Christian et al., 2009; Coyle et al., 1995; Health and Safety Executive, 2003; Johnson, 2007; Neal & Griffin, 2006). Specific management dimensions such as genuine commitment to safety and support of followers are tied to safety outcomes (Nahrgan et al., 2011; Spreitzer & Mishra, 1999). Although essential as a starting point to developing an understanding of culture, these studies provide a one-dimensional view that does not take into consideration the multidimensional nature of culture.

The sharp divide between the functionalist and interpretive approach to studying culture has limited both the maturation of safety culture theory and the research tied to organizational performance in high reliability and accident analysis. There is a need to marry the functionalist and interpretive approaches to culture in a way that considers the multidimensional aspects of leadership when designing research. The case study in Chapter 3 provides an integrated design and methodological structure that expands our understanding of the relationship between leadership behaviors, work team engagement, and safety outcomes in a high-risk environment.

3 Case Study Exploring Leadership, Work Team Engagement, and Safety Performance in a High-Risk Work Environment

This case study is situated in a large research and development laboratory that supports over 4000 employees.

The case study explores how the independent variables of high hazard work, historical safety performance, and staff perception of engagement may be mediated by leadership. The study provides a detailed contextual analysis of high-risk work teams. The high-risk work teams were derived from a statistical algorithm that predicts future incidents based on engagement and operational history (Caldwell & Larmey, 2012). Specifically, two groups of high-risk work teams were identified based on either a high or low number of predicted future incidents.

3.1 CONCEPTUAL FRAMEWORK

The study focused on the phenomenon of leader–work team behaviors that produce engagement and promote organizational reliability in a high-risk research environment. The theories and concepts associated with safety culture, high reliability, and authentic leadership that informed the conceptual framework of this study were discussed in detail in Chapter 2 and are summarized in the following paragraphs.

Concepts associated with safety culture and high reliability argue that leadership plays a pivotal role in developing an organization's capacity to anticipate problems by interacting with staff to detect weak signals of failure and taking deliberate steps to understand potential systemic factors contributing to incidents (Weick & Sutcliffe, 2007; Weick et al., 1999). The idea is that careful, mindful organizational practices can make up for inevitable limitations to the rationality of individual members (Weick & Sutcliffe, 2001). Leaders practice mindfulness by connecting with frontline staff in order to obtain the bigger picture of operations and being attentive to even seemingly minor or trivial signals that may indicate potential problem areas within the organization (Weick & Sutcliff, 2007; Weick et al., 1999).

Work team engagement is a key component for enhancing organizational reliability and safety performance (Frankel et al., 2006). The authentic leadership literature suggests that work team engagement is manifested as satisfaction, mutual trust, and respect (Avolio et al., 2009). When viewed from the context of engagement, safety culture, high reliability, and authentic leadership share similar concepts. Some of these concepts have been operationalized into leadership behaviors. The behaviors relevant to creating engagement and reducing the potential for a catastrophic failure include encouraging a questioning attitude, actively seeking out challenges and opposing viewpoints, and openly communicating (DOE, 2011; Hopkins, 2010; INPO, 2013; Weick & Sutcliffe, 2007).

The following case study uses the lens of engagement to apply the concepts of safety culture, high reliability, and authentic leadership at a research and development laboratory. The results of the study provide a better understanding of the leadership characteristics that support engagement and ultimately the reliable performance of work teams operating in a high-risk environment.

3.2 DATA COLLECTION

Data were collected using a mixed method concurrent triangulation design that addresses the multiple dimensions of culture (Creswell & Plano Clark, 2011; Terrell, 2011). The design employed two concurrent data collection phases (quantitative and qualitative) with equal priority given to each approach. Data were collected qualitatively through open-ended questionnaire responses, interviews, and participant observation. Quantitative information was extracted from the statistical analysis of survey data provided by work team members.

The primary purpose of the triangulation strategy is to expand our understanding of the phenomenon by converging or confirming findings within a single study (Terrell, 2011). Quantitative statistical results were directly compared and contrasted with qualitative findings to validate or expand quantitative results with qualitative data (Creswell & Plano Clark, 2011). This approach painted a richer picture of culture and supports Haukelid's (2008) assertion that fieldwork and observation are especially important to understand tacit knowledge and basic assumptions.

Tashakkori and Teddlie (2008) described mixed method studies as products of the pragmatist paradigm and suggest a pragmatic worldview is a way for researchers to shed light on mixed approaches and the quantitative–qualitative dualisms debated by purists. The pragmatic worldview regards knowledge as being constructed and based on the reality of the world that we experience and live in where we are constantly adapting to new situations and environments. Pragmatists prefer action to philosophizing and endorse theory that informs practice (Johnson & Onwuegbuzie, 2004; Creswell & Plano Clark, 2011). The pragmatic approach considers the interrelationship of how work is done, basic organizational assumptions, and patterns of activity within the organization that provide the context for behavior (Janićijević, 2011).

Multiple sources were used to collect data. Methods included semi-structured individual interviews, completion of a work team questionnaire, and participant observation. The methods were chosen such that each source of information can be used to inform the other.

3.2.1 Semi-Structured Individual Interview

Semi-structured interviews were used as a means of providing data to qualitatively explore leadership's description of and experience with engagement of self and others. Interviews were conducted and analyzed using a modified long interview technique (McCracken, 1988). Long interviews are concerned with cultural categories and shared meanings that provide a deeper understanding of how culture mediates action (McCracken, 1988). The interview questions were developed through lines of inquiry that are based on a review of the literature that forms the research question (McCracken, 1988). The interview protocol is described in more detail in Chapter 5.

3.2.2 Work Team Questionnaire

A work team questionnaire was developed and tailored specifically for the case study. Similar survey instruments were evaluated and a pool of potential questions was created and refined (Barling et al., 2002; Bass & Avolio, 1990; Jung & Avolio, 2000; Weick & Sutcliffe, 2007). Both open-ended and closed-ended questions were asked. Open-ended questions were used to provide more in-depth information pertaining to participants' experiences and viewpoints. The closed questions were placed into theoretical constructs based on behaviors that represent the construct (Crocker & Algina, 1986). A construct is an "unobservable or latent concept that the researcher can define in conceptual terms but cannot be directly measured…or measured without error. A construct can be defined in varying degrees of specificity, ranging from quite narrow concepts to more complex or abstract concepts, such as intelligence or emotions" (Hair et al., 2006, p. 707).

Many psychological and organizational traits, such as safety culture, are not directly observable or directly measurable and must be measured indirectly (Pedhauzer & Schmelkin, 1991). The questionnaire was intended to provide information on team member perceptions regarding leadership and how it has impacted the work team's attitude and behaviors. Perceptions are the way people organize and interpret their sensory input, or what they see and hear, and call it reality. Perceptions give meaning to a person's environment and make sense of the world. Perceptions are important because people's behaviors are based on their perception of reality. Therefore, employees' perceptions of their organizations become the basis on which they behave while at work (Erickson, 2013).

The work team questionnaire consisted of 10 statements on leadership and work team behavior that was scored on a 5-point Likert scale ranging from "not at all" (score 1) to "frequently if not always" (score 5). In addition, two open-ended questions asked the participant to describe leadership behaviors associated with their feelings of engagement.

Cronbach's alpha was used to measure the internal consistency or reliability of the cumulative data for the open-ended Likert scale survey questions using an accepted reliability level of 0.70 (Bland & Altman, 1997). Seven of the survey questions were placed into three constructs of collaboration, inclusiveness, and empowerment. The strength of their relationships was measured using Pearson's *r*. The Pearson correlation coefficient values were calculated between pairs of questions having the same dimension value.

3.2.3 Participant Observation

Participant observation provided the advantage of delivering data about participants that is beyond the conscious understanding of the respondents and may be reflective of underlying values and assumptions (Guest et al., 2013). It is interactive and unstructured. The data generated was dynamic, typically consisting of handwritten notes that were converted into field notes as soon as possible after the observation. The field notes contained more detail including a description of the observation context and the people involved, including their behavior and non-verbal communication. Participant observation protocol is described in more detail in Chapter 5.

3.2.4 Sample Selection of High-Risk Work Teams

The case study examined work teams and their leaders that operate in high-risk work environments. A model that predicts future incidents was used to identify high-risk work teams through work previously conducted by the author (Caldwell & Larmey, 2012). Chapter 4 describes the background and use of the predictive model.

For three years, profiles were created for high-risk work teams within the organization under study. The profiles contained basic information on the factors presumed to be connected to the risk for future accidents. Two of the factors used to predict the number of future incidents were staff engagement as determined by engagement survey and operational experience as determined by the number of incidents over the past three years (Caldwell & Larmey, 2012). Nine high-risk work teams were studied. Two of the high-risk work teams had a history of a low number of predicted future incidents and the remaining seven work teams had a history of a high number of future predicted incidents. The nine work teams were a representative sample of the high-risk population. The lower number of work teams in the former group was a reflection of the total small number of high-risk workgroups identified as having a low number of future predicted incidents. In addition, 23 associated supervisors, mid-level leaders, and senior leaders were also invited to participate in the case study.

3.3 DATA ANALYSIS

Data were analyzed using a concurrent triangulation strategy. The data analysis began with the qualitative analysis of work team open-ended questionnaire responses and individual leadership interview transcripts to provide a rich description of their espoused beliefs regarding the characteristics of leadership and engagement. Qualitative data were summarized and patterns, relationships, and themes were identified. Thematic analysis was performed through the phased process of coding to create established, meaningful patterns. The process included familiarization with data, generating initial codes, developing themes from codes, reviewing themes, and defining and naming themes that were crucial to the research question (Braun & Clarke, 2006).

Following the qualitative analysis, a quantitative analysis of the team members' responses to the 10 closed survey questions was performed. The quantitative description was synthesized with the qualitative data to understand statistics associated with survey results using an ethnostatistical approach (Gephart, 1988, 1993, 2006). Ethnostatistics analysis rejects the bifurcation of qualitative and quantitative data and posits that researchers need to understand both statistics and rhetoric for a thorough analysis of the data of any study. Ethnostatistics addresses *sensemaking* practices and tacit knowledge as part of the application of statistics. This study applied first-order ethnostatistics by investigating the activities, meanings, and contexts involved in producing the variables and statistics (Gephart, 1988).

Aligning with Schein's (2004) model of cultural assessment, results were presented from two perspectives: (1) leadership and engagement espoused characteristics and (2) team member and leader behaviors. These two perspectives facilitated the data analysis. Figure 3.1 illustrates the process for the analysis and interpretation of results using a concurrent triangulation strategy.

3.4 RESULTS

The results are presented in three parts. Part 1 describes the qualitative results. Part 2 presents the quantitative results. Part 3 provides a synthesis of qualitative and quantitative results, including findings. Nine work teams were chosen that were characterized as demonstrating a high or low probability for future incidents based

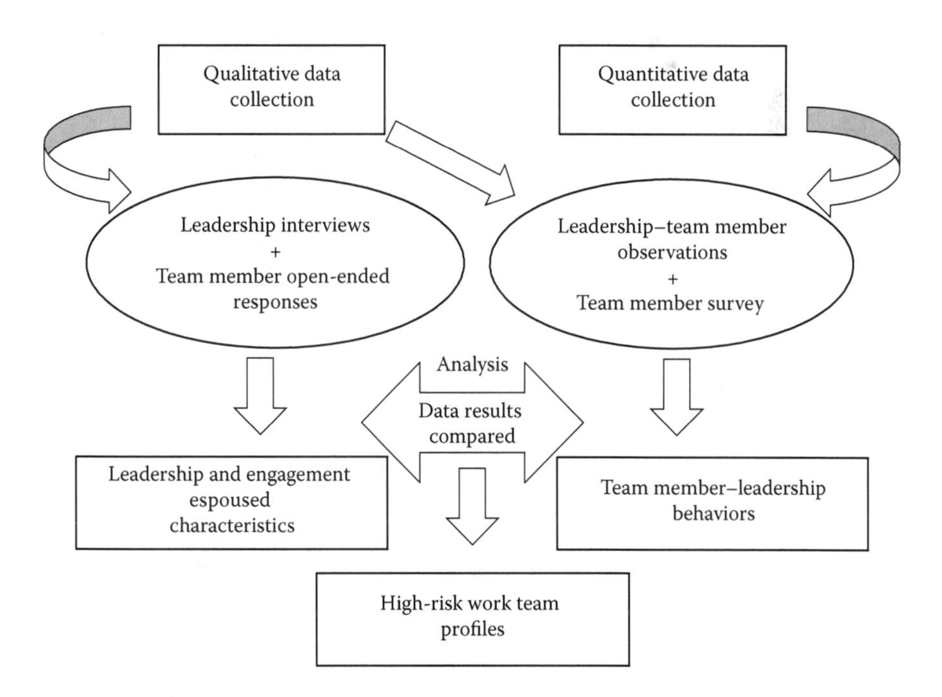

FIGURE 3.1 Concurrent triangulation strategy.

on a three-year history of incidents and scores on engagement and safety culture surveys. The nine work teams performed two distinct functions within the laboratory: research and support. The research function is tasked to acquire and execute project work within the schedule and budget. The support function is tasked to provide maintenance and safety assistance to the research function. Five of the nine work teams represented the research function and the remaining four work teams represented the support function.

Table 3.1 lists work team composition by function and predicted incidents.

In addition to the work teams, leadership associated with these work teams participated in the study. Twenty-three leaders participated and represented the nine work teams. Fourteen of the participants were first-line and mid-level leaders and nine of the participants were senior and executive-level leaders. The population of leaders interviewed represented approximately 85% of the available population. Table 3.2 lists the organizational function and level of the leaders interviewed.

3.4.1 Part 1: Qualitative Results

The first phase of qualitative analysis combined the data collected from two sources to develop themes on the leaders' and team members' perceptions of the espoused characteristics of leadership and engagement. The themes represented perceptions on the desired qualities of engagement based on personal experience.

Data were collected from two sources:

1. Interview of 23 leadership participants who were asked to describe their experience with engagement and how engagement has influenced their current behaviors, values, and beliefs in interactions with their work teams. Specifically, leadership described engagement factors important to them and specific experiences that support these factors.
2. Responses from 86 work team members to two open-ended questions that asked participants to describe specific motivators and de-motivators for their engagement at work.

TABLE 3.1
Work Team Participants by Function and Predicted Incidents

High-Risk Work Team	Function	Predicted Incidents
A	Support	High
B	Support	High
C	Support	High
D	Research	Low
E	Research	High
F	Support	High
G	Research	High
H	Research	Low
I	Research	High

TABLE 3.2
Leadership Interview Participants by Function and Level

		Level	
Interview Participants	**Function**	**Line/Mid**	**Senior/Executive**
1	Support		X
2	Support	X	
3	Support	X	
4	Support	X	
5	Support	X	
6	Support	X	
7	Research	X	
8	Support		X
9	Research	X	
10	Research	X	
11	Research	X	
12	Support		X
13	Research		X
14	Support	X	
15	Support		X
16	Support	X	
17	Support		X
18	Research		X
19	Support		X
20	Research	X	
21	Research	X	
22	Research	X	
23	Research		X

The leadership transcripts and team member responses were reviewed multiple times in an iterative process that identified codes, refined and attributed meaning to the codes, and then grouped the codes into themes. Themes were established based on the quality, frequency, and distribution of references. The first round of transcript review resulted in 61 first-order codes. Twenty-three of the first-order codes described the behaviors and characteristics of leaders associated with staff engagement such as accountability, backing-up staff, and honesty. The remaining 38 first-order codes described leader experiences with engagement such as demonstrating caring, acknowledging accomplishments, and listening to feedback. In the second round of review, second-order codes were identified through a process of clarifying and combining significant and similar aspects of the 61 first-order codes. The 12 second-order codes are listed in Table 3.3

The final third round reviewed the 12 second-order codes to further refine and clarify meaning. Five distinct but not mutually exclusive themes emerged. These themes described the participants' perception of the most valuable characteristics of leadership and engagement based on their personal experience. The themes were consistently expressed by most participants regardless of leadership level or

TABLE 3.3
Second-Order Codes

Code
Accountability
Encouraging teamwork
Freedom
Inspirational goals
Involved in work
Acknowledging accomplishments
Manager support
Ownership
Manager participation
Trust
Mentor or coach
Teamwork

TABLE 3.4
Categories, Themes, and Frequency of Unique Sources Associated with the Espoused Characteristics of Leadership and Engagement

Category	Theme	Unique Sources
Contributing to team members' ownership of their work	Leadership recognizes staff capability, gives them autonomy to accomplish work, and backs their decisions.	46
Nurturing teamwork	Leadership develops teams by encouraging collaboration and developing a team spirit.	39
Addressing performance	Leadership sets behavioral expectations, addresses performance issues, and provides positive reinforcement	32
Building relationships	Leadership demonstrates genuine care for team members by taking time to know them both as a person and a staff member and providing opportunities for development and mentoring.	29
Tying into the vision and mission	Leadership creates a vision, aligns goals, and communicates how staff contribute.	21

function. Table 3.4 lists the categories, themes, and frequency of unique sources (leaders and team members) associated with the espoused characteristics of leadership and engagement.

Due to the qualitative nature of this part of the study, data are presented in the form of participant quotes and summaries of their responses.

3.4.1.1 Theme 1: Contributing to Team Member Ownership of Work

The first and most prevalent theme that emerged as a characteristic of engagement was leadership contributing to team members' ownership of their work.

A key aspect of this theme is leadership understanding and recognizing staff capability. One team member observed feeling engaged when "covering refueling activities in a power plant" after being told by supervision, "You're the tech, handle it." Other team members cited more involvement in the work "when [leadership] got me engaged in the job planning and preparation" or when "management allowed me to be more involved in the decision making and up front planning." One staff member recalled, "[I felt disengaged when] I was not included in the planning of the work and was not included in the data work up or even saw the results to know if they were good or bad. I was ignored and shoved aside during data acquisition that I was actively involved in." Similarly, another team member stated, "When leadership dictates how a job will be I lose motivation, work less efficiently, feel as though my skills are underutilized. I cannot be creative, I am awfully reluctant to give feedback or make suggestions."

Another key aspect of this theme is being given the autonomy to accomplish work. Leader #2 recalled a personal experience of being given autonomy that impacted her early career: "It was just that freedom to go try something, go do it. I just kept [my manager] in the loop. So, I really appreciated him." Similarly, reflecting on an inspirational person earlier in their career, leader #13 observed that she was inspired to think outside of the box and to take ownership for a job "because there were no boundaries." Leader #20 recalled working with an engaged team that was "the best team that I worked with, the people were highly engaged in the work. I think that was driven by their empowerment. They could influence the outcome of the work. They felt like they were owners of their piece of work. They weren't just here is this piece of work, go execute it." With regard to her personal philosophy, leader #5 observed that "to motivate and continue with a highly engaged team is to give them more freedom within the constraints that we have to work within."

Many work team member responses also resonated with the notion of freedom and autonomy. One team member responded that he/she felt engaged when "leadership involves staff in the work and gives them autonomy to accomplish the work." Another stated, "They let me do my job." Other examples include: "For the most part, I was given autonomy in the work. The desired outcomes were well defined but how to get to the end result was up to me"; "[Leadership] allowed me to work independently and work out issues that arise."

A critical aspect of encouraging freedom and autonomy was leadership support when things went wrong. One team member recalled feeling engaged when her "manager stood back to let me make decisions and backed up those decisions." Another team member commented, "I feel the most attentive, absorbed, and involved when I also feel that my managers or group leaders have my back and are willing to stand up to a client or other manager regarding technical disputes, budget issues, or other issues that affect my work." Similarly, another team member observed, "Leaders were willing to share the risk and that allowed me to push the envelope in attempting to accomplish the project." Leader #11 remarked that when leadership "has your back it gives people courage to take on new paths and kind of not just sit around and wait. People are willing to take a lot of risk and good risk and kind of put themselves out there if they think… they've got a net to fall in." Leader #2 recalled an influential leader that "encouraged me to go do, make some changes. I always knew that he was there to backstop me. But he gave me some free rein. I always knew that if something didn't go quite right, that he would be there to backstop me. If I ever needed him, he was always there for me."

3.4.1.2 Theme 2: Nurturing Teamwork

Leadership nurtures teamwork by encouraging collaboration and developing a team spirit. Team members attributed engagement with leaders that "give credit to work well done and acknowledge the team" and leaders that take "time to discuss project objectives and take an interest in daily work and things that impacted daily activities." One team member recalled a time when "leadership instilled a great sense of team spirit in our team. As such, even if the task was rigorous and required working long weeks, I knew we were working towards a worthwhile goal and that I was working with people who are competent and who I can trust." Commenting on the attributes of teamwork, leader #13 stated, "We knew each other's strengths and weaknesses, and we knew when to play on them and we knew when to say I need help or let me give you help. I think that was what made our team so good."

Leader #11 described a leader that engaged a team by "encouraging of a lot of teamwork and encouraging … a lot of interaction between ourselves to make mistakes in front of each other, to back each other up, to share information." Leader #7 observed, "I'm a huge proponent of teamwork and how basically if you work as a team …you get to celebrate as a team and when things don't work out too well you agonize as a team."

Leader #7 described the critical role that teamwork plays in their success: "So, all of us have a pretty good technical background but it's how you bring those technical folks together to really get the most out of a project for the client that you are serving. So, I guess in a nutshell it's the technical part but it's the teaming part that really pushes this over the top." Similarly, leader #18 commented, "You have to get the right chemistry in the team and underlying trust. Unless you have that you're dead in the water."

Senior leaders also provided insight on developing a leadership team. Leader #1 stated, "I think the thing that I have realized more about [the laboratory] than earlier in my career is the whole leadership team working together to build [trust]. Because what I do can either be reinforced or hurt by the people that I work for and the people that work for me." Similarly, leader #15 observed, "People that had, in the engaged group, a strong leadership team, a cohesive leadership team, speaking with one voice, working effectively together, modeling that. They worked well together. My disengaged workgroup? The management team was constantly at odds with one another. Openly, they would debate and argue at times unprofessionally." Leader #18 commented that, "I actually like meeting [with my leaders] as a group more frequently than one-on-one because I think that builds this cohesion. It begins to have the whole group rely on each other. So, if they have questions, they will often call each other because their level of trust has been built as a team."

3.4.1.3 Theme 3: Addressing Performance

Both leadership and team members agreed that leadership must attend to performance by setting behavioral expectations, addressing performance issues, and providing positive reinforcement.

Setting behavioral expectations was discussed in the context of creating accountability. Leader #19 reflected that "the leader needs to create clarity, needs to create accountability in terms of performance." Leader #12 described leadership modeling of expectations: "I definitely think that you have to set an example. You have to set the model, by which you would expect your staff... kind of the minimum standard you would expect your staff to hold up." Similarly, leader #1 stated, "I try to set examples of how I expect behaviors or try to set behavioral standards."

Accountability at the worker level was frequently described by both leadership and team members as everyone pulling their own weight. Leader #2 told her team, "We're not going to pick and choose what [work] we do. We have a system where it's first in, first out." Similarly, leader #14 recalled that "there are times when maybe some other crew wants to take a shortcut on something and we can't do that and I let them know." Leader #9 commented that "very dynamic excited energized driving teams really want to succeed and there are hard things that come with that, like you've got to get rid of the deadwood, they've got to see you do that. They've got to see that you know if you have a level of expectation and everybody expects you to rise to it." Leader #2 agreed: "We did some discipline on some people that weren't doing their job basically and the group was thankful that I did that. They don't care for the deadwood any more than anybody else does, because somebody is going to pick up that work."

Team members also voiced frustration with leaders that condoned poor work ethics "by letting those who do not wish to work get away with it because they [leadership] want to avoid any and all confrontation. It's much easier for them to take the path of least resistance and ask those they know will do something, than to ask those people who are going to piss and moan and argue about having to get off their butts to do something." Leader #1 recalled an experience with a disengaged team and concluded that "in the end I looked at that and I did not blame the individuals, but I think it was the former managers of that team that built that relationship, built that culture. Allowed them to behave in a way that was unacceptable for years, and rewarded them for it."

Leaders agreed that individual performance problems reflected on team performance and dynamics. Leader #9 commented that "if you don't take care of problem children that's a cancer that will grow and kill a team extremely fast." Leader #18 reflected on a disengaged team experience: "It really boiled down to a single individual who was at the core of the whole issue. I just had to move him out of the organization. I heard how one individual can poison a whole team and I experienced it firsthand. I had no idea that it would be that profound. It was instantaneously we all had [a] common mind-set, we were marching forward." Similarly, leader #3 noted, "It only takes one or two bad eggs to bring the whole team down. One bad egg can just change the whole dynamics of the team. Generally, when you move them or something changes, the whole team is right back to where they were."

Leaders providing positive reinforcement for a job well done were deemed just as important to engagement as addressing poor performance, especially by team members. Team member comments include the following: "The moment your contributions are not acknowledged or rewarded in some way, one can feel that nothing they do matters, you become unattached. I have been there before"; "[I feel engaged] when leadership gives credit to work well done and acknowledges the team"; "[I feel engaged when] leadership understood and implemented ideas and shared who came up with those ideas to others; recognition shared to others boosts morale"; "When you have been working on a project for multiple years and you get acknowledged for the hard work and dedication you have committed to the project, it makes it easier to be involved, absorbed, and interested in what we're doing"; "My manager is very good at acknowledging work well done. He also makes himself available if you need help. This makes for a very engaged and positive work environment."

Leaders provided examples of the constructive effect of recognition. Leader #10 recalled a project manager that "tries to bestow credit where credit is due and make sure that his project team is recognized for their efforts." Leader #8 described a special coin that she had made to recognize staff. "I had 100 made, I am the only one that gives it out and this is for operational excellence. I'll just pop in with a coin and a citation and a certificate in front of their peers, and let others know why they are getting it." Reminiscing on an earlier career experience, leader #1 recalled being recognized by the company vice president who said, "I think we all know who is responsible for this." "He brought me up, and … I was emotional…. That was really a big deal."

3.4.1.4 Theme 4: Building Relationships

Leaders demonstrating genuine care for team members build strong relationships. Leader #13 reflected that, "If people know that you care about them you can deal with any situation. If they believe that you care through daily interactions when that bad thing happens, they are there [to provide support] because you care about them." Leader #20 recalled a leader that inspired her because "he cared about the work and the people who were doing it." Leader #13 described caring in terms of spending time in the team members' work space. "It always amazes me how happy they [team members] seem that I'm coming out. They are like wow, somebody like you cares and I'm thinking oh me like I am just me." Leader #7 described caring as "being authentic with people and getting down and talking with people. Coming in the labs just for no other reason than to show interest in what you were doing." Leader #2 described an experience turning a disengaged work team into an engaged work team: "I would say they were disengaged and kind of doing their own thing, and didn't feel respected. So, that's where it was taking care of their needs, getting things out of the way, showing them that I did care about them."

Providing the opportunity for personal development is yet another aspect of caring that leadership and team members perceived to play an important role in engagement. Recalling an earlier career experience, leader #12 observed, "What really engaged me was [management's] desire to give me more and more responsibility. To give me a lot of experiences, to get me out there and try a bunch of things. I think that's kind of what excited me." Similarly, leader #9 recalled being given an assignment "that was a great stretch opportunity, I mean an enormous stretch opportunity. I appreciated [my leader's] confidence and his faith."

3.4.1.5 Theme 5: Tying into Vision and Mission

Leaders and team members that can tie themselves and their work into a higher vision and mission report being more highly engaged and are motivated to succeed. To build a connection, leadership creates a vision, aligns the team goals, and communicates how staff contribute to the bigger picture. Leader #12 remarked that her engagement was influenced by senior leaders "allowing me the ability to create a vision. [Then providing the support needed to] make it happen and really getting behind you and believing that you can reach the goals and meet the milestones that you set out for yourself." Leader #18 recalled working for a leader that "set lofty goals and he truly believed we could achieve them. So, he was somebody that you could really get behind and be motivated to work for." This leader viewed leadership as "setting that big vision and having the charisma to make others believe that you can get there." Similarly, leader #15 recalled an engaged workgroup that "had a common purpose and goals and outcomes. They were all working towards the same thing. They had a common challenge and goal that the leadership put in front of them to achieve something that brought them together."

Both leaders and team members commented on the importance of establishing a connection to a larger cause. Leader #12 commented, "Working with the lab I have always felt engaged about the mission, the science mission here. So, that helps motivate me internally." Similarly, leader #17 stated, "My motivation doesn't come from the individuals that I work for. My motivation, my inspiration has always come from the mission." Leader #8 noted that staff "need to feel valued in what they do and that what they do is tied to a larger outcome. Everybody here will tell you they view this as the strategic facility for the nation." Similarly, leader #9 described her experience with the characteristics of engaged work teams as "doing work that's challenging to them, that's impactful, that they feel like they are making a contribution and a difference and they feel like they are empowered to control their lives or to contribute to a bigger cause."

Team members had observations similar to leadership when describing their experience with goals and engagement: "We had very clear-cut goals, you are all working towards the mission of producing electricity and producing plutonium. So, those were kind of highly engaging goals"; "[I was engaged when my] leader defined clear goals and objectives for solving an important national priority"; "[I was engaged when] leadership believed in the program itself. They conveyed to us how important our work was for the greater good." On the other hand, one team member recalled feeling disengaged when "I didn't feel the project goal/outcome was of importance and didn't feel the work I was doing mattered. In part, [this] could be failure of leadership to convey the importance of [the] mission." Another team member reported feeling disengaged when the connection to the goals and mission are "too complicated and indirect."

3.4.1.6 Participant Observations

In addition to the interview, qualitative data were collected using observations. Ten observations were made of leaders and work team members in various settings. The process was facilitated by my familiarity and cultural awareness of the organization in general, but this also introduced the opportunity for bias so care was taken to avoid interpretation of the observations. Work team members and leaders were observed as they engaged in activities that would probably have occurred in much the same way without my presence. Most of the observations were one-time events, but in several cases I had the opportunity to observe the same individuals in different settings. The observations noted verbal behavior, physical behavior, personal space, and body language.

The observations were intended to supplement the results from data collected through interviews and surveys and were focused on getting an overall impression of leader interactions with peers and team members. Table 3.5 briefly describes the leader observations.

3.4.2 Part 2: Quantitative Results

A questionnaire was distributed to nine high-risk work teams. The questionnaire collected data on work team perceptions of leadership behavior and how it has influenced work team feelings and behavior through responses to 10 closed questions.

3.4.2.1 Survey Response Rate

The questionnaire was distributed to 148 high-risk work team members. Eighty-six questionnaires were completed for a 58% response rate. The response rate fell short of the predetermined overall goal of a 60% response rate determined as part of the data collection protocol. The potential for non-response bias was evaluated by comparing response rates across work teams comprising the target population (Table 3.6). With the exception of work team I, the response rates appeared to be representative of the overall sample population. The reasons for the high non-response rate for work team I could not be determined. The work team was treated similarly to the other work teams except that the questionnaire was distributed close to a holiday. All of the responses were included in the quantitative analysis.

3.4.2.2 Reliability of Instrument and Construct Analysis

The overall Cronbach alpha result was 0.77, indicating an acceptable internal consistency for the survey on the rescaled data. Seven of the ten questionnaire items were placed into three constructs of collaboration, inclusiveness, and empowerment (Table 3.7).

The strength of their relationships was measured using Pearson's r. The Pearson correlation coefficient values were calculated between pairs of questionnaire items with the same dimension value. The comparisons for the constructs of empowerment and inclusiveness found a strong positive relationship between the items examined and the constructs. The construct of collaboration indicated a weak positive relationship. The remaining three items (2a, 2b, and 4e) were not included in constructs and were examined as single entities. Table 3.8 provides the results of the Pearson's r calculation.

TABLE 3.5
Leader Observations by Type and Description

Participants	Observation Type	Brief Description of Observation
Leader 6	Discussion	A miscommunication had affected the leader's ability to support one of his/her work team members who was investigated. The leader was quiet, non-judgmental, and offered suggestions for improving communications.
Leader 1	Interface meeting	The leader was quiet, but probing to understand emerging issues and reach agreement on the need to work together to better improve communications.
Leader 13	Interface meeting	There was a lot of peer-to-peer interaction on current issues. The leader was open, emphasizing creating partnerships and critical self-assessment.
Leaders 1, 15, 18, 19, 23	Senior leadership meeting	Safety and operational issues were discussed. Each of the leaders was supportive, encouraging, and questioning. They acknowledged accomplishments and how tough it is to keep a focus on preventing injuries and accidents.
Leader 4 and work team	Staff meeting	The leader shared information and encouraged discussion and the team members were quiet but attentive. When the team members were given the opportunity to speak, several issues surfaced. At this point the team members expressed frustration over the lack of communication. Although there was a detailed discussion on the issues and potential solutions, the leader also gave the impression that the solutions were out of their control. Through body language and comments, the work team appeared exasperated with the lack of input into issues that they are involved in and avenues for resolution.
Leaders 1 and 3	Safety interface meeting	The leaders were supportive of each other. The senior leader #1, although soft-spoken and amiable, openly challenged the group to probe deeper into issues and actively define what that organization believes "safe looks like." At that point, a defensive posture was taken by the leader's peers and the discussion became defensive with a decisive change in body language.
Leader 5 and work team	Staff meeting	The group of team members was attentive to the information exchanged by the leader. Many questions were asked and there was general interest in the topics discussed.

(Continued)

TABLE 3.5 (CONTINUED)
Leader Observations by Type and Description

Participants	Observation Type	Brief Description of Observation
Leaders 1, 3	Safety interface meeting	With the exception of the leaders being observed, the participants were quiet with most checking cell phones or working on their computer. There was little discussion.
Leader 11 and team members	Staff meeting and luncheon	Thirty-five team members attended the meeting hosted by the leader. The leader presented information, introduced new team members, and updated the group on current initiatives. Discussion was encouraged. There was much discussion and interest, and many questions.
Leader 6 and work team	Staff meeting	The leader schedules weekly meetings for those team members that are on shift that day and interfaces with all team members on a six-week cycle. The group of three was interested and attentive and appeared to feel comfortable raising questions. The leader relayed current information and encouraged discussion.

TABLE 3.6
Response Rate by Work Team

Work Team	Surveys Sent	Surveys Completed	Response Rate (%)
A	16	10	63
B	8	6	75
C	8	6	75
D	35	21	60
E	10	5	50
F	6	4	67
G	16	9	56
H	35	23	66
I	14	2	14
Total	148	86	58

3.4.2.3 Means Comparison

Overall means and standard deviations were calculated for questionnaire items and collectively for the three constructs and are displayed in Tables 3.9 and 3.10.

Each of the overall construct mean values (collaboration, inclusiveness, and empowerment) were greater than 4 on the 5-point Likert scale, indicating that work team members generally agreed that the characteristics associated with the construct

TABLE 3.7
Survey Items and Associated Constructs

Number	Item	Construct
2e	My leader understands things that could go wrong with our project or work assignments	Collaboration
2d	My leader listens carefully to different points of view before coming to conclusions	Inclusiveness
2c	My leader encourages open and honest debate	Inclusiveness
4b	My work team members including myself are involved in decisions that directly affect our work	Inclusiveness
4d	My work team members including myself often feel pressures that lead us to cut corners	Empowerment
4e	My work team members including myself understand things that could go wrong with our project or work assignments	Collaboration
4a	My work team members including myself are criticized when we report information that could interrupt work	Empowerment

TABLE 3.8
Pearson's *r* Calculation

Paired Items	Pearson's r
Collaboration (2e and 4e):	0.294[a]
Empowerment (4d and 4a):	0.488[b]
Inclusiveness (2d and 2c):	0.698[b]
Inclusiveness (2d and 4b):	0.521[b]
Inclusiveness (2c and 4b):	0.468[b]

Note: Means have been rescaled for questions 4b, 4a, and 4d.
[a] Weak positive relationship.
[b] Strong positive relationship.

are displayed in the work environment. Of the three remaining items not included in the constructs, two had overall means of less than 3:

2a My leader actively seeks out bad news.
4c My work team members including myself are rewarded for identifying potential problems.

These results indicate that work team members generally disagreed that the characteristics associated with these questions are displayed in the work environment.

Significant differences between means for the two groups of high-risk work teams were analyzed using a *t*-test. The group of work teams with high predicted future incidents was compared with the group of work teams with low predicted future

TABLE 3.9
Means for Questionnaire Items by Question

Questionnaire Item	Mean (SD)
2a. My Leader: Actively seeks out bad news	2.31 (1.21)
2b. My Leader: Discourages questioning	4.21 (1.05)
2c. My Leader: Encourages open and honest debate	4.21 (0.98)
2d. My Leader: Listens carefully to different points of view before coming to conclusions	4.06 (0.92)
2e. My Leader: Understands things that could go wrong with our project or work assignments	4.21 (0.98)
4a. My work team members including myself: Are criticized when we report information that could interrupt work	4.36 (0.93)
4b. My work team members including myself: Are involved in decisions that directly affect our work	3.77 (1.10)
4c. My work team members including myself: Are rewarded for identifying potential problems	2.80 (1.09)
4d. My work team members including myself: Often feel pressures that lead us to cut corners	3.90 (1.04)
4e. My work team members including myself: Understand things that could go wrong with our project or work assignments	4.34 (0.73)

Note: Means have been rescaled for questions 2a, 4a, and 4d.
SD = Standard deviation

TABLE 3.10
Means for Questionnaire Constructs

Construct	Mean (SD)
Collaboration:	4.27 (0.69)
Inclusiveness:	4.01 (0.85)
Empowerment:	4.13 (0.84)

Note: Means have been rescaled for questions 2a, 4a, and 4d.
SD = Standard deviation

incidents. For the dimension of inclusiveness, the mean value of work teams with low predicted incidents was significantly different from that of work teams with high predicted incidents. For the dimensions of collaboration and empowerment, there was weak evidence of a difference in means. Table 3.11 displays the *t*-test analysis of high-risk work team groups with high and low predicted incidents.

To provide supplemental information, the Tukey Honest Significant Difference (HSD) test (alpha level of 0.05) was calculated in conjunction with an Analysis of Variance (ANOVA) test to identify means that are significantly different from each other. Tukey's HSD makes multiple simultaneous mean comparisons by comparing

TABLE 3.11
***T*-test Analysis of High-Risk Work Team Groups with High and Low Predicted Incidents**

	Work Team Mean (Standard Deviation)	
Question/Construct	**Low Predicted Incidents $n=44$**	**High Predicted Incidents $n=42$**
2a. My Leader: Actively seeks out bad news	2.43 (1.15)	2.19 (1.27)
2b. My Leader: Discourages questioning	4.23 (1.08)	4.19 (1.04)
4c. My work team members including myself: Are rewarded for identifying potential problems	2.93 (1.15)	2.67 (1.03)
Collaboration (2e and 4e)	4.40* (0.60)	4.14* (0.77)
Empowerment (4d and 4a)	4.30* (0.78)	3.95* (0.90)
Inclusiveness (2d, 2c, and 4b)	4.23**(0.65)	3.78** (0.96)

*p<.10; **p<.05.

all of the means between groups. The Tukey HSD results comparing nine high-risk work teams are shown in Table 3.12.

For the construct of inclusiveness, the mean response for work team A was significantly different than the mean responses for all other work teams. This result is important because the significantly lower mean result in the dimension of inclusiveness for work team A primarily influenced the statistical difference in means between the high and low predicted incident work team groups shown in Table 3.11.

The lower values for work team A responses to the questionnaire cannot be explained with certainty. It is most likely that the results were influenced by several factors. First, the work team has experienced a large number of leadership and team member changes over the past three years. It is probable that engaged behaviors were not adequately established or reinforced by the new work team leaders. Signs of this were observed during leader–team member interactions during the timeframe that the questionnaire was completed. The observation noted visible frustration among team members over follow-up to specific concerns. This frustration also could have influenced the lower values assigned to the questionnaire. Based on this analysis, the responses were retained in the study and attributed to a more poorly engaged work team relative to the other work teams with high predicted incidents. More analysis is needed to understand the influence of organizational changes on the engagement of a high-risk work team.

3.4.3 Part 3: Data Synthesis and Comparison

Previous sections provided the results of qualitative and quantitative analysis. In this section, the results representing the espoused characteristics of leaders and current behaviors of leaders were further synthesized. Specifically, the themes representing

TABLE 3.12
Tukey HSD Results Comparison of Nine High-Risk Work Teams

	Work Team Mean (Standard Deviation)								
Dimension/Question	**I**	**D**	**C**	**E**	**F**	**G**	**H**	**A**	**B**
N	2	21	6	5	4	9	23	10	6
2a. My Leader: Actively seeks out bad news	3.00 (1.41)	2.76 (1.22)	2.33 (1.75)	2.00 (1.00)	2.50 (1.29)	2.11 (1.17)	2.13 (1.01)	1.70 (1.34)	2.67 (1.21)
2b. My Leader: Discourages questioning	4.50 (0.71)	4.19 (0.93)	4.17 (1.60)	4.60 (0.55)	4.25 (0.96)	4.33 (0.71)	4.26 (1.21)	3.60 (1.35)	4.50 (0.55)
4c. My work team members including myself: Are rewarded for identifying potential problems	3.00 (1.41)	2.67 (0.97)	3.00 (1.41)	3.40 (0.55)	2.00 (0.82)	2.67 (1.12)	3.17 (1.27)	2.20 (1.41)	2.83 (0.75)
Collaboration (2e and 4e)	4.50 (0.71)	4.40 (0.52)	3.83 (0.93)	4.40 (0.82)	4.38 (0.95)	4.22 (0.75)	4.39 (0.67)	3.70 (0.71)	4.58 0.38)
Empowerment (4d and 4a)	3.75 (1.77)	4.17 (0.93)	4.17 (0.52)	4.40 (0.22)	3.88 (0.85)	4.28 (0.57)	4.41[a] (0.62)	3.20 (1.77)	4.25 (0.76)
Inclusiveness (2d, 2c, and 4b)	4.50 (0.24)	4.16 (0.66)	4.06 (0.61)	4.27 (0.76)	3.92 (0.33)	4.22 (0.58)	4.30 (0.63)	2.60[b] (0.24)	4.06 (0.25)

Note: The variation in sample size is due to the variation in the number of team members that elected to complete the questionnaire.

[a] Work team H is significantly different from work team A.

[b] Work team A is significantly different from all other groups.

$\alpha = 0.05$.

espoused behaviors of both leaders and work team members were compared with indicators of the current state of the organization depicted by responses to Likert scale survey items and participant observations. Each theme was tied to related questionnaire items and/or constructs.

Team member responses to Likert scale survey items were designated as positive or negative based on the response mean. Taking into consideration central tendency bias, a mean value of ≥4.0 was considered positive and <4.0 was considered negative. Observations of leadership behavior were designated as either positive, negative, or not applicable based on the general demonstration of the aspects of each theme and the opportunity for those aspects to be demonstrated across the population.

Table 3.13 shows the general relationship between the espoused themes and the perceived current state for high-risk work teams and Table 3.14 compares the relationship between the espoused themes and the perceived current state for work team groups with low and high predicted incidents.

TABLE 3.13
Relationship between the Espoused Themes and Perceived Current State for High-Risk Work Teams

Theme	Team Member Perception		Leader Observation
Leadership recognizes staff capability, gives them autonomy to accomplish work and backs their decisions.	2b	+	N/A
	Collaboration	+	
	Empowerment	+	
	Inclusiveness	+	
Leadership develops teams by encouraging collaboration and developing a team spirit.	2a	–	Overall positive
	2b	+	
	Collaboration	+	
	Empowerment	+	
	Inclusiveness	+	
Leadership sets behavioral expectations, addresses performance issues, and provides positive reinforcement.	4c	–	N/A
Leadership demonstrates genuine care for team members by taking time to know them both as a person and a staff member and providing opportunities for development and mentoring.	2b	+	Overall positive
	Empowerment	+	
Leadership creates a vision, aligns goals, and communicates how staff contribute.	Not applicable		Overall negative

Note: + = mean value ≥4.0. – = mean value <4.0.
N/A = not applicable. There was no opportunity for observation.
Overall positive observation indicates that demonstrated aspects were generally characteristic of the theme.
Overall negative observation indicates that demonstrated aspects were generally not characteristic of the theme.

TABLE 3.14
Relationship between the Espoused Themes and Perceived Current State for Work Team Groups with Low and High Predicted Incidents

Theme	Low Predicted Incidents Team Member Perception		Low Predicted Incidents Leader Observations	High Predicted Incidents Team Member Perception		High Predicted Incidents Leader Observations
Leadership recognizes staff capability, gives them autonomy to accomplish work, and backs their decisions.	2b Collaboration Empowerment Inclusiveness	+ + + +	N/A	2b Collaboration Empowerment Inclusiveness	+ + – –	N/A
Leadership develops teams by encouraging collaboration and developing a team spirit.	2a 2b Collaboration Empowerment Inclusiveness	– + + + +	Overall positive	2a 2b Collaboration Empowerment Inclusiveness	– + + – –	Overall negative
Leadership sets behavioral expectations, addresses performance issues, and provides positive reinforcement.	4c	–	N/A	4c	–	N/A
Leadership demonstrates genuine care for team members by taking time to know them both as a person and a staff member and providing opportunities for development and mentoring.	2b Empowerment	+ +	Overall positive	2b Empowerment	+ –	Overall negative
Leadership creates a vision, aligns goals, and communicates how staff contribute.	Not applicable		Overall positive	Not applicable		Overall negative

Note: + = mean value ≥4.0. – = mean value <4.0.
N/A = not applicable. There was no opportunity for observation.
Overall positive observation indicates that demonstrated aspects were generally characteristic of the theme.
Overall negative observation indicates that demonstrated aspects were generally not characteristic of the theme.

3.4.4 Findings

Four findings emerged from the synthesis of quantitative and qualitative data:

> Finding 1: Behaviors associated with leaders contributing to team members' ownership of work, leaders nurturing teamwork, and leaders building relationships with work team members were generally consistently demonstrated for the leaders of high-risk work teams.

Finding 2: Espoused behaviors associated with leaders addressing performance and tying into vision and mission were generally inconsistently demonstrated or inconclusive based on a lack of data for the leaders of high-risk work teams.
Finding 3: Leaders of high-risk work teams were not perceived to actively seek out bad news from work team members.
Finding 4: Leaders of high-risk work teams with a low number of predicted incidents more frequently demonstrated resilient behaviors when compared to leaders of high-risk work teams with a high number of predicted incidents.

3.5 DISCUSSION

This section discusses the meaning of the findings and their relationship to previous theory and research. The findings address the behaviors of leaders of high-risk teams as well as the differences between high-risk teams with low and high predicted incidents. Specifically, the discussion will focus on how the findings relate to the theories and related concepts of safety culture, high reliability theory, and authentic leadership, including the supporting theory of psychological safety. Table 3.15 lists the findings and their nature.

TABLE 3.15
Description and Nature of Findings

Finding	Nature of Finding
Behaviors associated with leaders contributing to team members' ownership of work, leaders nurturing teamwork, and leaders building relationships with work team members were generally demonstrated for the leaders of high-risk work teams.	General finding associated with leaders of high-risk work teams
Espoused behaviors associated with leaders addressing performance and tying into vision and mission were inconsistently demonstrated or inconclusive based on a lack of data for the leaders of high-risk work teams.	General finding associated with leaders of high-risk work teams
Leaders of high-risk work teams were not perceived to actively seek out bad news from work team members.	Finding associated with a specific leadership behavior not expressed or demonstrated by leaders of high-risk work teams
Leaders of high-risk work teams with a low number of predicted incidents more frequently demonstrated resilient behaviors when compared to leaders of high-risk work teams with a high number of predicted incidents.	Findings from a comparison of behaviors of leaders of two groups of high-risk work teams

3.5.1 Demonstrated Behaviors of Leadership in High-Risk Work Teams

The analysis of qualitative and quantitative data on the characteristics of leadership and engagement generally converged around three themes. The themes were described as leaders contributing to team members' ownership of work, leaders nurturing teamwork, and leaders building relationships with work team members. It should be noted that most but not all of the work teams demonstrated traits associated with these three themes.

3.5.1.1 Team Members' Ownership of Work

The cognitive model of participative effects proposes that workers who participate in workplace decision making are informed by a richer, higher quality pool of information. Driscoll (1978) summarized that participation in decision making positively influences individual and organizational satisfaction provided that individuals have the right skill set, feel empowered to affect outcomes, and have the support of their leader. The affective model of participative effects proposes that participation in decision making will meet higher order needs such as respect and independence and make workers more satisfied with their jobs (Driscoll, 1978). A meta-analytic review of the effects of participation in decision making found support for both the participative and affective models and concluded that participation in decision making has an effect on both worker satisfaction and productivity (Miller & Monge, 1986).

These two models are not mutually exclusive and were both supported by the comments of leader and team member participants. Many of the study participants credited leaders with contributing to staff members' individual ownership of work by involving them in decision making, job planning, and preparation. These perceptions were reinforced by survey responses that supported the notion that the leader and team are actively involved in work.

Autonomy refers to the freedom to accomplish work including decision making and work processes (Nahrgang et al., 2011). Many team members resonated with the notion of freedom and autonomy to make decisions contributing to their individual engagement in accomplishing work. These sentiments align with safety culture traits endorsed by the nuclear industry and the Department of Energy (DOE, 2011; INPO, 2013). In addition, the organizational literature supports the notion that when leaders trust employees and give them decision-making abilities, employees develop more trust in their organization. Often leaders are hesitant to involve employees because they fear losing control. The very act of involving employees makes managers vulnerable by relinquishing their authority (Spreitzer & Mishra, 1999).

A meta-analytical study conducted by Nahrgang et al. (2011) found that a supportive environment included leadership support, social support, and organizational support. Leadership support was also perceived by staff members as a critical aspect of owning their work, especially when things go wrong. Leadership support also includes showing concern for subordinates and valuing their contributions and is discussed in the following paragraphs in relation to nurturing teamwork and building relationships.

3.5.1.2 Leaders Nurturing Teamwork

The concept of psychological capital posits that personal recognition reinforces goals and creates a positive motivational state that is based on the team members' sense of goal-directed energy (Luthans et al., 2015). This study supported the value of recognition and teamwork. Recognition was observed to take many forms such as celebration, walking the hallways, saying "thank-you," and personal involvement in what is important to others. Team members credited teamwork with leaders that acknowledge the team's accomplishments and participant observations found that many leaders nurtured teamwork through formal gatherings to celebrate as a team. Similarly, the nuclear industry and the Department of Energy emphasize the reward and recognition of positive performance as attributes of a strong safety culture (DOE, 2011; INPO, 2013). Leaders providing positive reinforcement for a job well done was deemed just as important to engagement as addressing poor performance, especially by team members.

Other aspects of teamwork were also supported by the current study. Survey responses for the construct of inclusiveness generally supported the notion that team members and leaders are involved in work decisions; each voice is heard and all viewpoints are considered before making a decision. Participant observations found that the interactions among leaders and team members were supportive and encouraged discussion and information exchange. In addition, participant observation generally found that leadership interaction with peers focused on establishing an open partnership.

Team members frequently identified trust as the outcome of nurturing teamwork. Trust enabled the team to weather difficult situations and feel safe to rely on each other. Authentic leadership builds trust through a value-based leadership approach that creates an inclusive and caring climate (Avolio et al., 2004). Trust is built through relational transparency, which is characterized by communicating openly and being real in relationships with others (Gardner et al., 2005). Authentic leaders build teams characterized by clear elevating goals, competent team members, unified commitment, and a collaborative work climate (Gardner et al., 2005).

3.5.1.3 Leaders Building Relationships with Work Team Members

Strong relationships build a team's resilience to detect, bounce back from, and cope with challenges (Edmondson, 2012; Weick et al., 1999). Relationship building with team members was generally attributed to availability and face time with leaders. The majority of leaders interviewed indicated that they interact with staff on a daily basis. Many of the preferred interactions were described as face-to-face during walk arounds to socialize, share information, and better understand the issues and risks associated with work. Participant observation found that some leaders arranged their schedule to spend impromptu time with staff, although some leaders relied on team members approaching them rather than seeking and initiating interaction. Survey responses in the dimension of inclusiveness generally supported the notion that team members and leaders are involved in work decisions; each voice is heard and all viewpoints are considered before making a decision. These behaviors were also supported through the leader and team member perceptions that were previously discussed.

In the context of developing psychological capital, authentic leadership posits that mentoring relationships as well as attention, recognition, and positive feedback on performance inspire individuals with confidence to succeed at challenging tasks (Luthans et al., 2015). Leaders that genuinely acknowledge and recognize contributions inspire both individual and team confidence. Recognizing potential, providing the opportunity for personal development, and providing mentoring is yet another aspect of caring that leadership and team members perceived to play an important role in engagement. The more complimentary the relationships within the team, the more likely the group will have a shared belief in its capabilities (Edmondson, 2012). Bandura (2000) used the term *collective efficacy* to predict the level of group performance. Bandura (2000) proposed that the stronger the belief a team holds about their collective capabilities, the more they achieve.

In addition, leaders acting as teachers and mentors were perceived to be influential in developing future leaders. One team member described an early career experience leading high-risk activities under the mentorship of her manager: "My manager would listen to issues and recommend solutions in a teaching rather than lecturing manner. I felt empowered and accomplished when the work was completed. My manager's approach resonated with me and I have adopted his techniques when leading teams." Another team member commented that she was positively impacted by leaders that "were always willing to let me grow and give new challenges." These sentiments contrast with a team member's comment that "working with a manager that doesn't pass along new job or stretch assignment opportunities is very disheartening." An executive leader recalled the relationship with his/her leader as "more of a coaching peer… it certainly made it easier to participate as a peer with them rather [than using a] top down command and control [approach]." Similarly, leader #8 observed the development of a mentoring relationship with one of his leaders: "He allows you to go off and tackle the issue, keeping him informed. At that point, you've built trust among you and that person and roles tend to morph into one of collaborative mentorship."

3.5.2 Unconfirmed Behaviors of Leadership in High-Risk Work Teams

The results of the analysis of qualitative and quantitative data were generally inconclusive with respect to two themes. The themes were described as leaders addressing performance and leaders connecting team members to a vision and mission. It would not be expected to observe explicit examples of leaders addressing inadequate performance; however, it would be expected to observe leaders connecting to the organizational vision, mission, and goals. It should be noted that most but not all of the work teams did not demonstrate or weakly demonstrated aspects of the latter theme. These themes were supported by attributes of safety culture, high reliability theory, and authentic leadership, but the behaviors were either inconsistently demonstrated or the data were lacking.

Finding 2: Espoused behaviors associated with leaders addressing performance and tying into vision and mission were generally inconsistently demonstrated or inconclusive based on a lack of data for the leaders of high-risk work teams.

3.5.2.1 Addressing Performance

Setting boundaries, allowing team members to feel a sense of ownership, and holding them accountable establishes an environment of psychological safety and increases team members' ability to collaborate, learn, and innovate (Edmondson, 1999). The conditions for maintaining a psychologically safe work environment were discussed by the study participants within the context of setting expectations and accountability for performance.

Many of the leaders interviewed reflected that clear expectations, accountability in terms of performance, and modeling of expectations were necessary leadership behaviors for staff engagement. A small number of participants voiced frustration with non-confrontational leaders that condoned poor work ethics.

Study participants agreed that individual performance problems reflected on team performance and dynamics. This sentiment supports Edmondson (2012) who reflected that individuals who believe in the competency and responsibility of all team members are more trusting and cooperative. When negative consequences occur, the leader must provide the rest of the team with justification for the difficult action to protect them from the fear of receiving a similar punishment. This maintains the group's psychological safety and supports a just culture (Edmondson, 2012).

While both leaders and team members agreed that leadership must attend to performance by setting behavior expectations and addressing performance issues, perceptions concerning these remarks were not corroborated with other sources. Although the participant comments align well with the safety culture and psychological safety literature, more information is needed to understand the extent of these behaviors within the organization.

3.5.2.2 Tying into Vision and Mission

Authentic leaders build trust through relational transparency by sharing goals and values with their team (Gardner et al., 2005). In terms of psychological capital, the hopeful manager possesses goals that excite others, goal-directed willpower, energy, and determination. They set the context for followers to determine their own goals and stretch their limits (Luthans et al., 2015). These concepts were supported by one observation where the leader clearly tied the successful work of the work team to a higher mission and team members that reported being motivated by the example set by their leaders.

When authentic leaders openly share their motive for pursuing organizational goals, followers understand how their work functions fit into the big picture (Gardner et al., 2005). Both leaders and team members commented on the importance of establishing and understanding their connection to a larger cause. A senior leader reflected, "I can help translate that [team members] are part of something bigger. Whether they are grounds crew that are cleaning up and maintaining our campus, or they are craft that are turning a wrench or maintaining the nightshift power operations, that they contribute to something bigger than themselves."

Feedback from leaders and work team members indicated that creating a vision, aligning goals, and communicating how staff contribute is an important behavior in the organization; however, these remarks were not corroborated with other data collected. The significance of establishing a leader–follower connection to vision and mission was an unexpected result of the study that was not addressed in the literature reviewed in Chapter 2. Although the participant comments align well with the authentic leadership literature, more information is needed to understand the extent of these behaviors within the organization.

3.5.2.3 Actively Seeking Bad News

Work team member responses to the statement, "My leader actively seeks out bad news" indicated that the behavior was not generally supported in the work environment. This finding represents a subtle but important difference in leaders encouraging the reporting of issues (which was seen to some extent) and actively seeking negative information. This behavior was not brought up during leadership interviews or team member descriptions of engaged behaviors.

Finding 3: Leaders of high-risk work teams were not perceived to actively seek out bad news from work team members.

Survey responses to the statement, "My leader actively seeks out bad news" had a mean value of 2.31. The relatively low mean value indicates that team members generally perceived that leaders did not assertively look for problems. Team member perception in this area may imply that leaders are uncomfortable or unwilling to listen to and act on information that might be openly challenging shared assumptions. Leader reluctance could in turn translate to team member reluctance to communicate bad news upward. Team member perception in this area may also indicate an environment that is low in psychological safety, which decreases the likelihood of learning behaviors such as discussion and the reporting of errors (Edmondson, 2012). Similarly, survey responses to the statement, "My work team members including myself are rewarded for identifying potential problems" had a mean value of 2.80. The relatively low mean value may indicate that team members generally perceive that identifying potential problems is not valued by leadership.

More information is needed to understand the extent of condition and context behind the survey responses. Establishing an environment that provides team members with the safety to assert an unpopular opinion to leaders is important as it allows for innovation and adjustments that could avert larger problems.

3.5.3 Comparison of High-Risk Workgroups Based on Potential for Future Incidents

Finding 4: Leaders of high-risk work teams with a low number of predicted incidents more frequently demonstrated resilient behaviors when compared to leaders of high-risk work teams with a high number of predicted incidents.

The leaders and team members of the work teams with low predicted incidents distinguished themselves from the group with high predicted incidents in several aspects. These qualities were either absent or noticeably less demonstrated by the leaders and work team members associated with high predicted incidents. Three distinguishing qualities appeared to support a higher level of resilience, which enables the organization to detect and bounce back from error.

1. Collective Mindfulness: Collective mindfulness is a foundational concept of resilience that is characterized by a state of constant checking for subtle indications of failure within high hazard environments (Weick & Roberts, 1993). The term collective mindfulness is used to describe the combined interactions of numerous individual activities that form a collective mental process (Weick & Roberts, 1993). Roberts and Bea (2001b) believed that the collective thoughts of a team create a synergy that can improve decision making and influence members to contribute ideas and actions that facilitate group performance. The notion of collective mindfulness was supported by work teams with lower predicted incidents in several ways.
   - The leaders associated with work teams having low predicted incidents each had a diverse background in terms of roles previously held within the organization. Interviews indicated that their varied experiences increased their understanding of the needs of the organization and enabled them to work effectively with their leadership, peers, and followers. A diverse background is especially helpful to varying diverse points of view.
   - During interviews with the leaders of work teams with low predicted incidents, teamwork and trust were emphasized almost to the exclusion of individual accomplishment. Teamwork was described at many levels including senior leadership teams, division leadership teams, workgroup teams, and project teams. One leader summed it up by saying, "It's just that team atmosphere and the respect, trust in the team that really, really is a differentiator between being good and great and that's just a philosophy that I've got that I've just learned as I've been here." Leader interviews stressed working together for the best interest of the organization. One leader observed that the leaders "are experienced and not self-serving, we get along extremely well and we play to the good of the whole division."
2. Voice: Establishing an environment that supports worker voice builds a greater sense of ownership for the organization and increases resilience. Leaders encourage voice through openness to non-conforming ideas and by encouraging creativity. Successful organizations must be particularly attentive as organizational success can promote denial and arrogance which will discourage voice (Luthans et al., 2015). This result is strongly supported by team members' perception that they and their leaders are involved in work decisions; each voice is heard and all viewpoints are considered before making a decision. The survey response mean in the dimension of inclusiveness reported by team members with low predicted incidents was statistically higher than the mean for the team member group with high predicted incidents.
3. Preoccupation with failure: Preoccupation with failure refers to the constant preoccupation with potential errors and failures (Weick & Sutcliffe, 2007). Organizations practicing preoccupation with failure use incidents and near misses as indicators of a system's health and reliability. Organizations that are preoccupied with failure systematically collect and analyze warning

signals and are attentive to even seemingly minor or trivial signals that may indicate potential problem areas within the organization (Weick & Sutcliffe, 2007). High-risk industries rely on learning from events that have a low impact on the organization but also touch on organizational concerns such as near misses and minor accidents (Lampel et al., 2009).

Encouraging disclosure of mistakes and signs of a breakdown in controls was valued by both leaders and team members of high-risk work teams with low predicted incidents. This trait was not demonstrated in the workgroups with high predicted incidents. For example, one of the leaders established a leadership advisory team with her senior staff. According to the leader, the purpose of the team was to solicit other points of view on how things are going with the work team, particularly in the context of safety. Another leader commented that "people take individual ownership and responsibility when they start seeing things breaking down to flag those early and bring some leadership attention to it." Leaders were also aware of establishing a psychologically safe environment to admit mistakes. One noted that "in safety programs if you beat people up every time they come and tell you they had an issue, they tend to not tell you …instead of rewarding people for coming forward and saying I made a mistake." Yet another leader stated, "It's okay to say I don't know. It is okay to say I saw wrong that I made mistakes. It's okay to be open and honest with folks and the staff aren't really looking for some manager up on some pedestal that never makes a mistake. They are looking for just real, honest, transparent dialogue."

3.6 RESEARCH LIMITATIONS

Two research limitations were identified. The first limitation was researcher bias: as the researcher, I have direct experience with the population participating in the research. To address this limitation, I was aware of the potential for researcher bias and intentionally endeavored to set aside my preconceived assumptions of the organization to best capture the essence of the phenomena being studied.

The second limitation was the inability of this study to be widely generalized due to the limited number of participants and the focus on one specific organization. Participant inclusion in the study was driven by the use of profiles that were generated from a statistical algorithm that identified high-risk work teams. The number of work teams identified was constrained by the identification of high-risk workgroups in the organization. Nine work teams representing over 80% of high-risk work teams in the organization were identified and invited to participate in the study. The overall sample size of high-risk work team participants ($n = 86$) provided statistics well within the acceptable range of the phenomena in question. The high-risk group consisting of nine teams was further split into two subgroups: high predicted future incidents (seven work teams) and low predicted future incidents (two work teams). The two work teams classified as having low predicted incidents produced a sample size of $n = 44$, which is 51% of the total sample. So, although the number of work teams associated with low predicted incidents was small, the total number of participants was representative. It is also important to note that the small number of teams selected is based on the small number of work teams that met the criteria using the predictive algorithm.

The nature of the study did not invalidate the results but additional research will need to be conducted to determine if the results can be duplicated within other industries.

3.7 METHODOLOGICAL CONSIDERATIONS

The study's research protocol provided robust assessment methods that captured a wide range of information and facilitated a structured exploration of the perceptions, behaviors, and espoused beliefs of the high-risk work teams and their leadership. The mixed method concurrent triangulation approach validated the convergence of some of the data results, but left some areas of the study wanting more information. The study may have benefited from using the interview and questionnaire information as a reference point to develop more specific data sources that would help triangulate results. Future work should consider an iterative approach to data collection that allows for additional data collection to validate emergent themes.

Work team member identification of the category of staff that best described their leader was variable. Although most participants referred to their leader as the individual in their direct organizational hierarchy, 41% of the participants identified an individual outside of the "chain of command." This included project managers, scientists, and engineers. According to the research protocol, the leader interviews were taken exclusively from the organizational hierarchy and potentially excluded a significant group of individuals that have the most influence on a team member's day-to-day activities. It is also possible that the questionnaire items may have been biased toward a more traditional leadership relationship and could have potentially missed the opportunity to collect different perspectives. This concern is partially offset by the varied experience of some leaders that were interviewed but should be considered for future studies.

3.8 IMPLICATIONS FOR THEORY, RESEARCH, AND PRACTICE

As discussed earlier in this chapter, many team members that participated in the study resonated with the notion of autonomy to make decisions contributing to their individual engagement in accomplishing work. Similarly, INPO (2013) emphasizes individual ownership in the preparation and accomplishment of work and encourages leaders to promote ownership and accountability. In the medical industry, *nurse autonomy* refers to the nurse's ability to assess a patient's needs. The value and contribution of nurse autonomy in improving nurse satisfaction and the quality and safety of patient outcomes have been consistently demonstrated (Weston, 2010). The concept of autonomy and its implications for high-risk work environments was not revealed during the literature review. Autonomy implies individualism and personal responsibility, which appears to align with high reliability performance principles that are based on the view that human intervention can prevent accidents but needs to be considered in the context of collective decision making. The role of autonomy is an area for future research. For example, the knowledge generated by the research can be used to further explore the relationship between autonomy and collective decision making in high-risk organizations.

The experience gained in completing the project may inspire other researchers to apply an innovative approach such as the predictive model for the selection of high-risk workgroups as a sample population. Inputs to the model focused on data that could provide information about work teams associated with three key variables: exposure to severe hazards, staff engagement, and past operational experience. These variables could be modified to address the needs of the research design. When used in conjunction with the concurrent triangulation mixed methods approach, the study further marries the science of predictive analysis and the art of organizational development.

This study also has practical implications for organizational learning. The results of the mixed method research design can be used to develop strategies to address leadership competencies to strengthen workplace engagement and increase resilience in high-risk work environments.

3.9 CONCLUSIONS

Complex accidents cannot be prevented through reliance on technology alone as the leadership behaviors associated with safety culture and high reliability theory in an authentic context must also be considered. The associated characteristics of each are inextricably connected and can be part of a leadership strategy to avert organizational failure.

Traditionally, high reliability organizations are a unique class of businesses where the paramount value that drives all decisions is safety. This case study adapted safety culture principles and the traditional high reliability organizational model to a loosely coupled, highly matrixed, multi-mission research and development environment. The results indicated that aspects of the models can be leveraged to influence organizational effectiveness outside the traditional model.

This case study augments the results of previous safety culture and high reliability research and provides a richer understanding of the complex relationship between leadership, work team engagement, and safety performance. Most importantly, this case study provides preliminary evidence that suggests that specific inclusive leadership behaviors may be effective in increasing resilience for high-risk work teams.

Section II

4 Creating the Capacity for Organizational Resilience

Traditional business models typically use a reductionist view of managing risk by keeping the management of risk separate and parallel with individual business functions. This model disregards how risks are affected by the interaction of people, processes, and the external environment and does not prioritize risk across the corporation.

Interactions introduce complexity into risk analysis. When studied individually, single component failure modes can be anticipated, but when the components are integrated into a larger system, unanticipated interactions can occur that lead to catastrophic events. Additionally, the mind-set that past performance predicts future outcomes does not fully consider the complexity of risk relationships throughout the hierarchy that could result in an operational failure.

Risk prioritization requires coordination at a high level in the organization. When risk is left to be managed at a lower level within an organization, risk is often conservatively treated with the same weight because the appetite for risk is low. For example, in the area of safety, managing lost work days does not give insight into the identification and mitigation of serious operational failure, yet in many organizations occupational safety is given priority. In this chapter, risk management techniques to increase organizational resilience will be discussed.

4.1 RESILIENCE

The term *resilience* has been described in the literature for over 30 years. Table 4.1 provides 27 definitions of resilience that are sorted into 9 categories based on meaning. Most frequently, resilience refers to either an adaptive outcome after an event or the capacity to adapt (Sonnet, 2016). Both of these definitions are well suited for use in theories of risk management as they are influenced by the non-linear and chance nature of events brought on by the complexity of the twenty-first century.

It is not surprising that when referring to the risk management of safety systems, Vogus and Sutcliffe (2007) offer the most relevant and comprehensive description that is directly tied to the concepts of reliability. Vogus and Sutcliffe (2007) describe resilience as being vitally prepared for adversity by having a collective reserve of capability that allows a team to make adjustments when challenged, rebound, and emerge stronger. Their description of resilience includes both the ability to course correct and the ability to rebound when challenged, often referred to as the latent and active aspects of resilience.

Resilience is not an attribute but rather a capability that is developed. Resilience is accomplished by growing a rich awareness of detail that could result in error detection, error correction, and avoidance of disaster. This assumes that individuals

TABLE 4.1
Definitions of Organizational Resilience

Category	Definition of Organizational Resilience	Author
1. Capacity, ability, or capability	a. For continuous reconstruction…requires innovation with respect to those organizational values, processes, and behaviors that systematically favor perpetuation over innovation	Hamel & Välikangas, 2003, p. 55
	b. To continue to operate and to provide goods, services, and employment critical to the ability of communities to be resilient	Lee et al., 2013, p. 35
	c. To effectively absorb, develop situation-specific responses to, and ultimately engage in transformative activities to capitalize on disruptive surprises that potentially threaten organization survival	Lengnick-Hall et al., 2011, p. 244
	d. To rebound or bounce back from adversity, conflict, failure, or even positive events, to progress, and to increase responsibility	Luthans et al., 2006, p. 28
	e. To expeditiously design strategies for the immediate situation	Mallak, 1997, p. 174
	f. For resisting, absorbing, and responding, even reinventing if required, in response to fast and/or disruptive change that cannot be avoided	McCann et al., 2009, p. 45
	g. To anticipate and manage risks before they become serious threats to the operation, as well as being able to survive situations in which the operation is compromised, such survival being due to the adequacy of the organization's response to that challenge.	McDonald, 2006, p. 173
	h. To self-renew over time through innovation	Reinmoeller and van Baardwijk, 2005, p. 61
	i. To survive and thrive in an environment of change and uncertainty	Stephenson et al., 2010, p. 3
	j. To make positive adjustments under challenging conditions and emerge strengthened and more resourceful	Vogus and Sutcliffe, 2007
	k. To survive an unscheduled disruption or major crisis through its adaptability using proven and integrated risk management, crisis management, and business continuity management processes	Teoh and Zadeh, 2013, p. 2
	l. To be vitally prepared for adversity; to have a broader store of capabilities	Vogus and Sutcliffe, 2007, p. 3418
	m. To cope with unanticipated dangers after they have become manifest, learning to bounce back; to manage surprise	Wildavsky, 1988, pp. 77 and 98

(*Continued*)

TABLE 4.1 (CONTINUED)
Definitions of Organizational Resilience

Category	Definition of Organizational Resilience	Author
	n. To keep, or recover quickly to, a stable state, allowing it to continue operations during and after a major mishap or in the presence of continuous, significant stresses	Wreathall in Hollnagel et al., 2007, p. 275
2. Function	a. A function of the overall vulnerability, situation awareness, and adaptive capacity of an organization in a complex, dynamic and interdependent system	McManus et al., 2008, p. 82
3. Attribute	a. A fundamental quality of individuals, groups, organizations, and systems as a whole to respond productively to significant change that disrupts the expected pattern of events without engaging in an extended period of regressive behavior	Horne and Orr, 1998, p. 37
	b. A multidimensional, organizational attribute that results from the interaction of three organizational properties: cognitive resilience, behavioral resilience, and contextual resilience	Lengnick-Hall and Beck, 2005, p. 750
4. Measure	a. The magnitude of disturbance the system can tolerate and still persist	Mamouni Limnios et al., 2012, p. 104
	b. A measure of the persistence of systems and of their ability to absorb change and disturbance and still maintain the same relationship between populations and state variables	Holling, 1973, p. 14
5. Phenomenon	a. A multidimensional, socio-technical phenomenon that addresses how people, as individuals or groups, manage uncertainty; a continuously moving target that contributes to performance during business as usual and crisis situations	Lee et al., 2013, pp. 29–30
6. Result	a. The result of relational reserves and the financial reserves to enable the maintenance of relational reserves	Gittell et al., 2006, p. 324
	b. The result of strategically managing human resources to create competencies among core employees, that when aggregated at the organizational level, make it possible for organizations to achieve the ability to respond in a resilient manner when they experience severe shocks	Lengnick-Hall et al., 2011, p. 243

(Continued)

TABLE 4.1 (CONTINUED)
Definitions of Organizational Resilience

Category	Definition of Organizational Resilience	Author
	c. When responses create negative-feedback loops that absorb jolts' impacts and loosen couplings between organizations and their environments... when responses expose new causal relationships that then modify theories of action, augment behavioral repertoires, and alter structural configurations and slack resource stockpiles	Meyer, 1982, p. 520
	d. The result of design structures that are resilient sources of collective sensemaking; resilient groups that are capable of four things: improvisation, wisdom, respectful interaction, and communication	Weick, 1993, p. 638; Weick, 1996, p. 145
7. Art	a. A kind of craft skill, or an artistic interpretation and response to singular, unexpected, anomalous events as opposed to a rationalized, predetermined response to what is regular or expected	Kendra & Wachtendorf, 2003, p. 45
8. Myth	A myth constructed politically to make sense of what happened; a way of retrospectively making sense of the radically surprising discovery of something entirely unknown by explicitly referring to the capacity to deal with rapid and radical change as well as having the capacity to survive and even benefit from this change.	Kuhlicke, 2013, p. 74
9. Puzzle	One of the great puzzles of human nature, like creativity or the religious instinct; a hot topic in business; a buzzword	Coutu, 2002, pp. 46–47

Source: Reprinted from *Employee Behaviors, Beliefs, and Collective Resilience: An Exploratory Study in Organizational Resilience Capacity* by M. Sonnet, 2016, PhD thesis, Fielding Graduate University. Copyright 2016 by Marie Sonnet.

possess a certain level of competence and understand the technical components of the business and also have diverse, integrated systems knowledge. Resilience enables organizational survival by providing the ability to cope with uncertainty and change over the long term.

Indeed, the concept of resilience has reinvented the field of risk management where safety considers the ability to succeed through the behaviors of people by recognizing impending change. Developing both individual and collective abilities for resilience allows the organization to adapt to changes and unplanned situations.

Why is resilience important? Resilience requires a new set of skills such as gathering seemingly disparate information, looking for the unexpected, recognizing opportunity, and learning from what goes right as well as what goes wrong. Retrospective analyses of the Deepwater Horizon and Bhopal disasters illustrate the consequences when organizational leadership discourages resilient behaviors.

In 2010, President Obama established the National Commission on the BP Deepwater Horizon Oil Spill and Offshore Drilling. The National Commission examined the relevant facts and circumstances concerning the root causes of the Deepwater Horizon explosion and found that

> "BP did not share important information with its contractors, or sometimes internally even with members of its own team. Contractors did not share important information with BP or each other. As a result, individuals often found themselves making critical decisions without a full appreciation for the context in which they were being made (or even without recognition that the decisions were critical)." (National Commission on the BP Deepwater Horizon Oil Spill and Offshore Drilling, 2011, p. 123)

Similar to the explosion on Deepwater Horizon, in the case of the Union Carbide Bhopal chemical disaster there were also signals that were not heeded. Union Carbide was one of the first multinationals to invest in India. India wanted to expand its crop productivity and be self-sufficient so decided to begin the manufacture of pesticides (Browning, 1993). The pesticide factory in Bhopal was started in 1969. A drought in India in 1982 and 1983 severely reduced pesticide sales. The drop in sales forced a reduction of 335 workers at the Bhopal plant. As skilled workers left they were replaced with less-educated workers. Maintenance was also reduced. Eventually, the plant's six safety systems designed to prevent a leak were either inoperable or failed due to management's belief that the risk of a leak was small when production was not ongoing.

For BP and Union Carbide, preoccupation with failure was not an organizational norm.

Overall, it was the accumulation of signals of failure that when combined resulted in both massive disasters. The organizational causes of both disasters are deeply rooted in the histories and cultures of their industries and the governance provided by the associated public regulatory agencies. In both cases, the values of the parent company reinforced "unreliable" and "unsafe" behaviors in their subsidiaries and subcontractors, which degraded the safety margins beyond recovery.

4.2 PRIORITIZING AND MANAGING RISK ACROSS THE ORGANIZATION

Successful leaders minimize the bad day events that can threaten the life expectancy of the enterprise by optimally managing risk. Managing risk increases resilience. The term "risk" is defined in a broad sense as the analysis of potential events, their consequence and likelihood, and the hazard or threat causing the event. Managing risk involves the ability to detect and minimize the impact of unforeseen events on the organization's assets. Assets that have exposure to loss include staff, property, business volume, profit, information, and reputation. Inherent risk describes exposure before intervention and residual risk refers to exposure after intervention.

Effectively managing the portfolio of organizational risk within a corporation requires a basic change in the way business is conducted and the organizational culture. Enterprise risk management (ERM) emerged in the late 1980s as an extension of hazard risk management and increased in popularity following September 11th and the Enron scandal as Chief Executive Officers and government leaders wanted to identify and manage new exposures. ERM examines risk holistically across an organization by identifying the major financial, business, and hazard risks of an organization, forecasting their significance in business processes and systematically addressing critical risks. When applied appropriately, ERM safeguards against a decision in one area unintentionally impacting overall organizational results (Hampton, 2009).

Once the specific risks that might threaten an organization are understood, they are treated and monitored. Integrating high reliability and a safety culture philosophy into an ERM framework provides an opportunity to understand more completely and focus on the critical risks that could impact the mission of a corporation by evaluating the most dangerous exposures and seeking early warning for risks that might otherwise be missed. The principles of ERM, safety culture, and high reliability are complimentary and when integrated provide a better understanding of how behavior influences risk. Figure 4.1 depicts the complementary relationship between ERM, safety culture, and high reliability.

How might the leadership of an organization that performs high-risk activities mitigate a bad day event that could threaten the organization's survival? Leadership must actively manage activity-based risk and behavior-based risk across organizational functions.

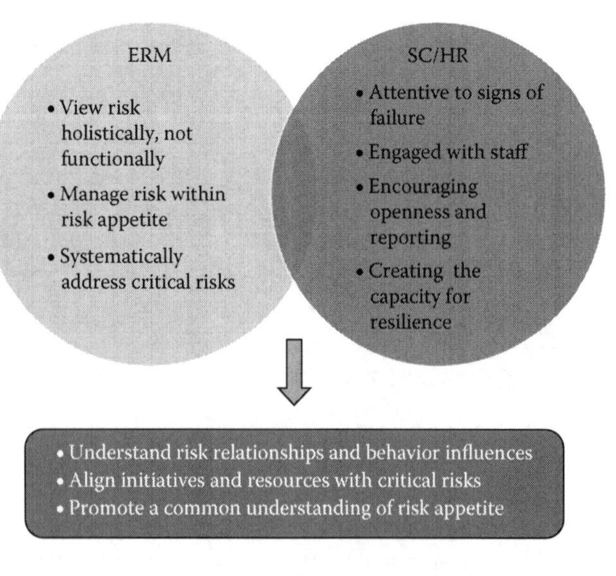

FIGURE 4.1 The complementary relationship between ERM, safety culture, and high reliability.

4.3 LEADERSHIP, RISK, AND PERFORMANCE

4.3.1 Activity-Based Risk

To begin understanding risk at the enterprise level, leadership should first consider the inherent risk that is intrinsic to the high-risk work activities being performed within the organization. The magnitude of inherent risk varies with the nature of work activity. Examples of bad day outcomes associated with inherently high-risk environment, safety, health, and security (ESHS) activities include

- Multiple injuries or death from the unsafe handling of radiological, chemical, or biological material
- Environmental release that impacts the community
- Property damage that destroys critical infrastructure and capability
- Significant infractions of state or federal regulations
- Loss of classified information
- Unsanctioned release of sensitive technology
- Cybersecurity event

The following pages describe a qualitative method for assessing activity-based risk using a tabletop format. The process involves a discussion among the stakeholders associated with specific ESHS programs and activities. To reduce subjectivity, persons with the greatest knowledge and experience in the functional area are chosen to provide input into the analysis. The analysis team makeup is cross functional to provide diverse viewpoints and a more nuanced picture of the total risk associated with a program. Typically, the group consists of management, a subject matter expert, a system user, and representatives of associated support functions. Although basically qualitative in nature, the discussion includes both qualitative and quantitative information to develop risk statements and implement controls. The process considers historical data, informed and expert opinion, and stakeholder needs. Information is gathered by asking a series of five questions:

1. What are the program objectives?
   The ESHS program objectives are typically characterized as providing protection or prevention, ensuring compliance, or enabling the organization's mission.
2. What are the risk statements that may prevent you from accomplishing your objectives?
   Risk statements are events tied to the risk source that have a negative impact on the organization.
3. What is the likelihood of occurrence and consequence for each risk statement assuming the program is functioning as intended?
   Likelihood is the probability or frequency of an occurrence and consequence describes the impact of the event occurring. After the likelihood and consequence are determined, the overall risk ranking is then reviewed by the group to validate that significant risks were captured and they appropriately reflect the operating environment.

Tables 4.2 and 4.3 provide examples of likelihood and consequence descriptions that can be used to assess risk events. Specific descriptions should be customized to meet your organization's needs and appetite for risk.

4. What are the controls used to mitigate the impact or frequency of an unwanted event as described in the risk statement?

   To focus the discussion, controls are described within three broad categories: people, processes, and engineered tools. Definitions and examples of controls are listed in Table 4.4

5. Which of the controls associated with the risk statement are essential to mitigating inherent risk and how well are they working?

   Essential means that if one of these controls were to fail or degrade it would significantly increase the potential for a bad day event.

When the process described above is performed across functions within the organization, it allows the organization to view the risk profile across the enterprise and focus on those higher likelihood/consequence risks that are more weakly controlled. Table 4.5 provides an example of an activity-based risk summary for a chemical safety program.

The ultimate challenge for leadership is balancing the mitigation of the inherent risk of high-risk activities with the organization's appetite for risk. The goal is to achieve and sustain an optimal level of acceptable risk for work. Since the consequences of an event have the potential to severely impact the organization, the level of acceptable risk should be acknowledged and approved by senior leaders in the organization. The organization's effectiveness at managing risk can be monitored through performance indicators for controls considered essential to managing the risk. Figure 4.2 illustrates the relationship between risk and performance.

Addressing activity-based risk and the controls associated with people, processes, and engineered tools described earlier is necessary but not sufficient to prevent operational incidents. Maintaining the reliability of technical systems is dynamic and is affected by the practices and behaviors of people. Leadership has the ability to set and reinforce performance expectations to develop resilience that considers

TABLE 4.2
Likelihood of the Risk Statement Occurring

Likelihood	Description
Highly unlikely	The event is very unlikely to occur in the next five years. This does not imply impossibility, merely high improbability; seldom happens, infrequent, rare, or has not happened before.
Unlikely	The likelihood of at least one occurrence during the next five years is much less than non-occurrence.
Possible	There is about a 50% chance of at least one occurrence during the next year.
Likely	The likelihood of at least one occurrence during the next year is much greater than non-occurrence.
Almost certain	Given no changes, the event is almost certain to occur at least once during the next year.

TABLE 4.3
Consequence if Risk Statement Occurs

Consequence Table						
Scale	**Mission Impact**	**Regulation and Image**	**Asset Loss**	**Project/Service Interrupt**	**Org Staff Injuries**	**Public Injuries**
Minimal	Little or no impact on the achievement of objectives or capability	Unsubstantiated, low impact, low profile, or no news items	Little or no impact on assets	< =1 week	No injuries	
Minor	May degrade the achievement of some objectives or capability Failure to meet key milestone (award fee goal/appraisal rating)	Substantiated, low impact, low news profile	Minor loss or damage to assets Equipment/information or facility damage, loss or theft less than $10K	> 1 week but < =1 month	Minor medical treatment	
Serious	Significant loss of award fee Max fines/penalties	Substantiated, public embarrassment, moderate impact, moderate news profile Project/client losses	Major damage to assets Equipment or facility damage > $10K and < $1M Equipment loss or theft > $10K and < $1 M	> 1 month but < = 6 months	Significant medical treatment	Medical monitoring/ testing of member(s) of the public
Disastrous	Significantly degrades the achievement of objectives or capability	Substantiated, public embarrassment, high impact, high news profile, third party actions	Significant loss of assets (>$1M) Some loss of vital or sensitive information	> 6 months but < = 1 year	Death of extensive injuries	Permanent disability, chronic or irreversible illness for member(s) of the public
Catastrophic	Significant capability loss and the achievement of objectives is unlikely	Substantiated, public embarrassment, very high multiple impacts, high widespread news profile, third party actions	Complete loss of assets Facility damage beyond habitable usage	> 1 year	Multiple deaths or severe permanent disabilities	Loss of public life

Note: Table 4.2 and Table 4.3 adapted from the University of Washington Enterprise Risk Management Toolkit 8/12/2010. Copyright 2007–2010 by the University of Washington.

TABLE 4.4
Definition and Example Controls Associated with People, Processes, and Engineered Tools

Control Category	Definition and Examples of Controls
People	Controls associated with people generally provide assurance that subject matter experts and workers have the appropriate knowledge, skills, and abilities to perform their function. E.g., selection and qualification of specialists, training of frontline workers, procedures that describe roles and responsibilities of staff performing critical functions, and committees and boards that oversee specific high-risk activities.
Processes	Controls associated with processes generally provide assurance that administrative systems provide checks and balances and are functioning as intended. E.g., procedures for performing hazardous work activities, procedures that limit quantities of hazardous materials, licenses and permits to perform work, signs, postings and barricades, and preventive maintenance of equipment.
Engineered tools	Controls associated with engineered tools generally assure equipment is functioning as intended and adequately maintained. E.g., locks and keys, physical barriers, ventilation systems, alarms, and interlocks.

TABLE 4.5
Example Activity-Based Risk Summary for a Chemical Safety Program

Element	Description
Chemical safety program objective	Safe and compliant program that protects staff, the public, and the environment
Risk statement	Deflagration or fire from improper treatment of chemicals causes injury or property damage
Likelihood and consequence	Unlikely/Serious
	Likely to have an interruption of services >1 month
Essential controls	Health and safety professional
	Procedures for handling, monitoring, and testing of chemicals
	Fume hoods, storage cabinets
	Training of chemical users

how behaviors interact with the processes and technology of the work environment. Behavior is driven by the culture of the organization. So, in addition to addressing risks that are associated with specific program activities, leaders must strategically manage risk for work teams most vulnerable to future incidents due to low engagement or poor operational performance.

4.3.2 Behavior-Based Risk

Integrating the high reliability philosophy into an enterprise risk framework provides an opportunity to more completely understand the full complement of risks

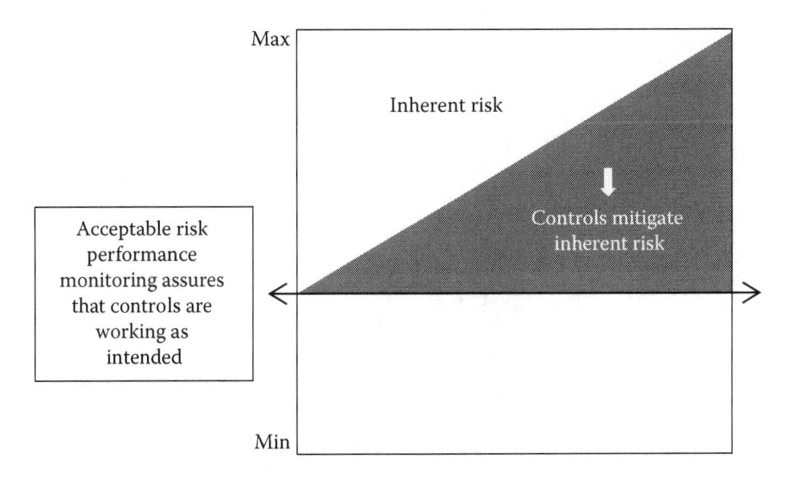

FIGURE 4.2 The relationship between risk and performance.

that could impact the mission of a corporation by focusing on behaviors and identifying early warnings for risks that might otherwise be missed. Predictive modeling is a method that facilitates the understanding of behavior-based risk by forecasting work teams at risk for a catastrophic event based on exposure to high-risk hazards, staff engagement, and their operating experience. The predictive model introduced in Chapter 3 was designed using the framework and basic assumptions of the job demands-resources (JD-R) model to identify work teams at the highest risk of future incidents (Bakker & Demerouti, 2007; Demerouti et al., 2001). The JD-R model proposes that job demands and resources are two sets of working conditions that can be found in every organizational context (Bakker & Demerouti, 2007; Demerouti et al., 2001). Job demands may be inherently negative, or they may turn into job stressors when meeting the demands requires substantial effort. Examples of job demands include high-risk activities, work pressure, and emotionally demanding interactions (Bakker & Demerouti, 2007; Demerouti et al., 2001). Job resources help individuals deal with job demands and have the potential to motivate them. Job resources include aspects of the job that help employees achieve work goals, reduce job demands, and stimulate personal growth and development. Examples of job resources include participation in decision making, leadership support, work autonomy, and a positive workplace climate (Demerouti et al., 2001). The predictive model assumes that organizational work teams most at risk for future incidents have the greatest exposure to high-risk hazards, the lowest levels of worker engagement, and the most problematic past operating experience.

Unlike approaches that associate accidents with specific tangible causes, the predictive model integrates the contributing aspects of employee engagement, safety culture, accident history, and exposure to high-risk activities to provide a unique lens through which to identify and manage safety risk across the organization. Depending on the available resources, risks, and needs of the organization, the rigor of predictive modelling used to forecast high-risk workgroups can be adjusted. A more

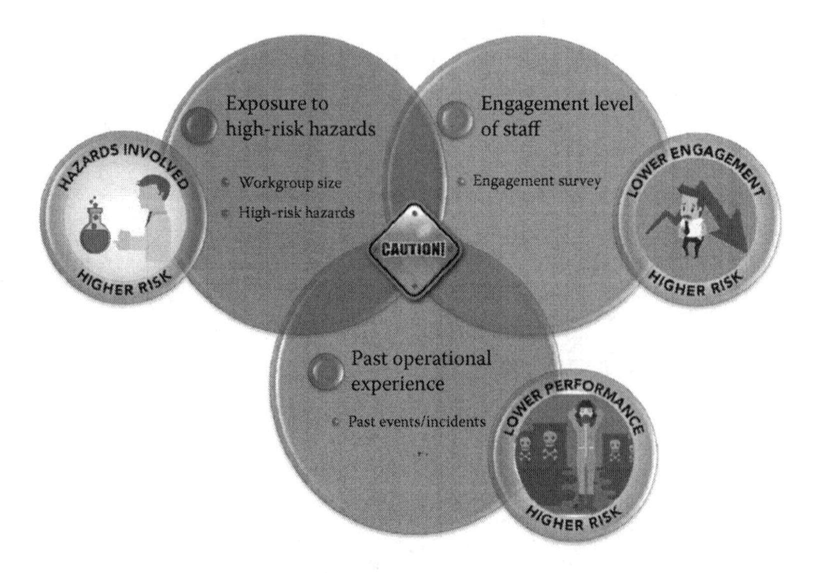

FIGURE 4.3 Conceptual model: Predicting future incidents for high-risk work teams.

rigorous statistically derived algorithm used in the case study (Chapter 3) can be applied to larger organizations with well-established data streams, or a less rigorous, qualitative, expert-based model approach can be applied to smaller organizations. A graphical depiction of the conceptual model is shown in Figure 4.3.

4.3.2.1 Applying a Predictive Model for ESHS

To begin the predictive process, a source of data should be identified for each of the areas that contribute to the model. Ideally, the data source should be easily sorted at the workgroup level and sustainable for long-term trending. Examples of data types and collection methods that might be used to better understand high-risk hazards, worker engagement, and workgroup operating experience in the context of ESHS follow.

The ESHS *high-risk hazards* chosen are those deemed most likely to cause a catastrophic outcome for the organization. Such outcomes might result in death or severe permanent disabilities, significant capability loss, damage to the organization's image, or a loss of assets. A brief list of examples of high-risk hazards might include work with certain classified information, radioactive materials, select agents, explosives, energized electrical equipment, peroxide formers, and foreign travel. After the hazards are identified, they are associated with individuals who work with high-risk hazards. Potential sources for collecting data on personnel working with high-risk hazards are training records, licenses, and certifications. Finally, exposure to high-risk hazards is quantified by summing the number of high-risk hazards at the work team level.

Survey data is the most useful data stream to provide insight on *staff engagement*. Consideration should be given to survey questions that are related to communications, trust, teamwork, and respect and questions should be tailored to meet the needs of your organization. Questions should be set up using a 5 or 7-point Likert scale format that ranges from a “strongly agree” sentiment to a “strongly disagree” sentiment.* The percentage for disagree or strongly disagree is calculated using the percentage of all questions in the survey for which members of the group responded either “disagree” or “strongly disagree.”

Operating experience considers the history of incidents or operational issues involving staff that may or may not have resulted in harm. The nature of incidents will vary depending on the function of the organization. Much of the literature has used informal or “recollected” data such as self-reports of behavioral safety and experts’ rating of safety level because of a lack of data. Examples of incidents associated with ESHS may include quality issues, security events, chemical spills, vehicle accidents, first aid cases, near misses, and procedure/regulatory issues. Potential sources for collecting data are non-conformance reports, logbooks, and hotlines.

Large data sets and complicated relationships are candidates for applying a statistical algorithm. The predictive model algorithm used in the case study in Chapter 3 was developed using negative binomial regression. The algorithm to calculate the predicted number of incidents in the following year for a given work team applied the additive effects of key predictor variables for each work team and converted them into a single outcome of incidents. Each of the inputs was transformed using various mathematical functions (e.g., log, square root) and then multiplied by a pre-calculated coefficient. The products were summed and added to a pre-calculated constant to generate the predicted number of incidents for a particular work team in the following year. A version of the resulting algorithm that calculates the predicted incidents is shown next (Caldwell et al., 2017):

$$\mathrm{Exp}\left(-1.2852 + \mathrm{A} + \mathrm{B} + \mathrm{C} + \mathrm{D}\right)$$

where:

A = 0.1932*log[(group size)*(laboratory incidents/person)]
B = 0.1353*sqrt(number of hazards)
C = 0.0247*(percent “disagree” or “strongly” disagree from engagement surveys)
D = 1.1757*log(1 + 3-year average of past incidents)

In addition to the predictor variables previously mentioned, the model takes into account the size of the workgroups. Group size was based on the number of direct reports in the group. Larger groups are expected to have more incidents due to having more people. Every person should have a non-zero probability of having an accident, so adding a person to a group should, on average, increase the group’s expected number of incidents.

Figure 4.4 is an example of work team summary information for predicting future incidents using a statistical algorithm.

* Developing and interpreting a questionnaire requires technical expertise. Chapter 5 discusses guidelines for developing questionnaires and provides sample questions associated with leadership.

Risk factor	Indicator	Workgroup	Org avg	Adjustment (model)
Size of workshop	Size of workshop	20 (24th percentile)	16	+1.2 incidents per yr
Exposure to high-risk hazards	# of work activities with high-risk hazards with which workgroup members are currently active (hazards include worker safety and health, environmental, and radiological)	65 (4th percentile)	8	+1.9 incidents per yr
Engagement level of staff	# of "disagree" or "strongly disagree" responses by workgroup members to questions on engagement survey	50% (1st percentile)	3.7%	+1.3 incidents per yr
Past operational experience	Avg # of incidents (e.g., injury/illness and vehicle accidents) per year in which workgroup members are involved Other factors modeled but not as significant contributors	7.3% (3rd percentile)	1.0	+4.6 incidents per yr
Expected # of incidents per year (based on model)	Expected # of incidents per year based on all factors	9.0 (1st percentile)	1.0	+9.0 incidents per yr (1.2+1.9+1.3+4.6)=9.0

Recommendations:

- Determine actions that can be taken to reduce risk
- Implement actions and monitor progress
- Share best practices and lessons learned

Notes:

High % disagree on many culture survey questions, including, "supervisor visiting workplace", concerns respected", and I am encouraged to report concerns".

FIGURE 4.4 Example work team summary information for predicting future events.

4.3.3 Understanding Performance

After the data are collected, it is analyzed to identify work teams at high-risk for future incidents. Using the model, at-risk work teams can be identified and shown graphically on a heat map (Figure 4.5).

The horizontal axis represents the additional number of incidents predicted based on exposure to high-risk hazards and workgroup size. The vertical axis represents the additional number of incidents predicted based on the past number of incidents and survey satisfaction scores. Two clusters of at-risk work teams can be identified based on either a high or low number of predicted future incidents. The work teams with a lower number of predicted incidents appear to be mitigating the risk through better engagement and a lower number of historical incidents. These work teams may be good candidates for benchmarking.

The data and analysis results can be useful to all levels of leadership within the organization. For example, the information may be used by senior and mid-level management to detect symptoms of a shift in organizational norms for what is acceptable and strategically reverse it. First-line managers can use the information for personal development to strengthen leadership behaviors and to engage their work teams.

Understanding behavioral performance and focusing on high-risk work teams is important to developing an adaptive capacity to manage the unexpected failures before they become a serious threat to the organization's operation. The question remains: how can a work team's resilience be cultivated to instinctively respond under challenging conditions?

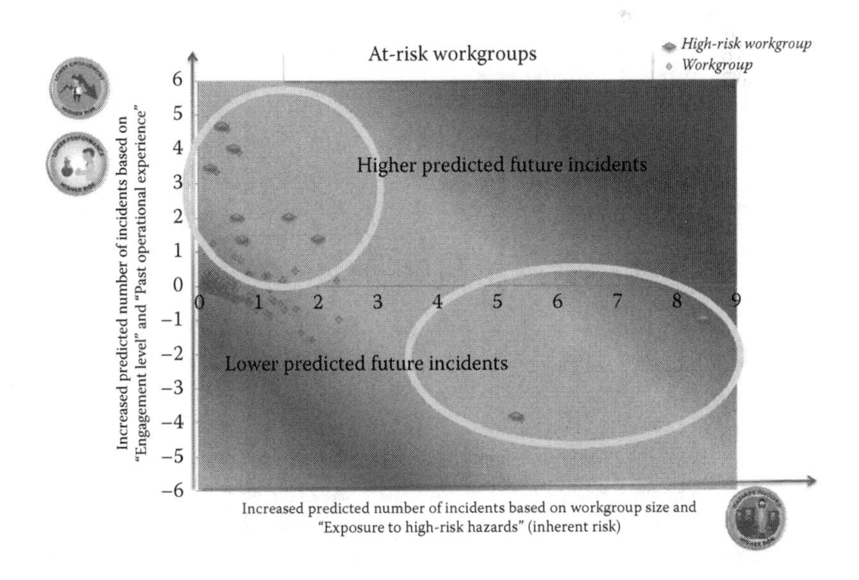

FIGURE 4.5 At-risk work teams with higher and lower predicted future incidents.

4.4 SENSEMAKING AND ENHANCING ORGANIZATIONAL RESILIENCE

Sensemaking is a collective group effort to comprehend complex events and is a key concept tied to resilience. Anticipatory resilience mechanisms stress the collective capacity of the group to compensate for individual weaknesses and are essential for cultivating sensemaking. Through the interaction of our individual cognitive processes, the actions we choose to take, and the reflection that is done in partnership with others, we form a shared reality based on which we take further action. Sensemaking improves organizational resilience by adjusting to demands and detecting and correcting unexpected failures. Low-probability, high-consequence events defy interpretation and impose severe demands on sensemaking. People think by acting, and to sort out a crisis requires an action that simultaneously generates a response that is used for sensemaking and also contributes to the unfolding crisis. The nuclear disaster at Fukushima Daiichi in March of 2011 provides some thought-provoking insights on sensemaking and how high-risk organizations might cultivate it.

On March 11, 2011, the largest earthquake in their recorded history struck Japan. The epicenter was approximately 180 km from Fukushima Daiichi. When the earthquake struck, the three operating, on-site, nuclear reactors at Fukushima Daiichi all automatically shut down or "scrammed." After a reactor scram, residual heat caused by the decay of fission products must be removed by cooling systems to prevent the fuel rods from overheating and failing (IAEA, 2011). Maintaining enough cooling to remove the decay heat in the reactor was the main priority for the nuclear plant workers as the events unfolded that Friday afternoon.

When the crisis struck, the workers first tried to make sense of the situation and looked for reasons to enable them to stay on the "normal" course. Their reasons were drawn from institutional training, expectations, and acceptable justifications. The operating crew of the plant expected that the system's design margin would mitigate and control the situation. When these design features did not control the situation as the workers expected, sensemaking helped them identify alternative actions.

In the first hours of the crisis, the workers' identities were challenged. They began to shift from highly trained nuclear technicians that monitored system parameters and worked to a strict set of procedures whose step-by-step actions have been carefully analyzed to soldiers facing unimaginable scenarios that required them to think and act independently with minimal communications and creatively employ every means necessary to prevent disaster. The designed defenses were defeated and the unfolding crisis was under the direct control of the human action of those working on shift that day.

To protect the integrity of the vessel and containment, the operators began preparations to vent steam to control the pressure and inject water to keep the fuel covered. The workers relied on their past experience and knowledge of the plant systems to address the dire situation since procedures did not exist for opening valves using batteries, compressors, and gas cylinders or for injecting water into the reactor core using fire engines (TEPCO, 2011).

Weick (1988) argues that by striving to make technology operator-proof, we move the dynamics of enactment to an earlier point in time where incomplete designs are enacted into unreliable technology. To complement technology, it is equally important to consider strengthening the anticipatory and resilience mechanisms used by workers. In the case of Fukushima, the accident response was enhanced by actively cultivating a collective mind-set built on institutional memory. Institutional memory and openness to possibilities allow individuals to see alternative possibilities and make connections. Resilience and reliability emphasize understanding beyond the rote memorization of procedures or the oversight of regulatory bodies. Workers do what is right based on collective knowledge that is applied to future experience. Leadership can provide the tools that create the capacity for resilience. On the other hand, regulations and procedural compliance are barriers to resilience: the question in a high-risk organization is how do we balance both?

The following actions can expand a workgroup's portfolio of organizational practices and patterns of behavior to prepare them to respond to the challenges of a bad day event:

1. Establish a fully developed network of resources to encourage the formation of long-term relationships and cross-functional collaborations that bridge traditional boundaries.
2. Provide rotational assignments to expand the integrated knowledge of each team member. Cross training provides flexibility and gives workers a better understanding of how each function fits together to accomplish work.
3. Implement tools and technologies that are designed to function seamlessly with the group's collective work function and facilitate resilient behaviors.

4.4.1 What's the Benefit?

Reliability is affected by the actions of people. Taking a risk-based approach that considers behaviors in addition to the technical aspects of controlling unwanted events provides a deeper understanding of the way that workers react to workplace demands. The approach enhances organizational resilience by

- Facilitating informed decision making
- Aligning organizational understanding of the drivers of risk
- Enabling early detection of subtle issues
- Providing performance feedback to leadership

4.5 SUMMARY

How can organizations become more resistant and resilient to operational upsets? They must change the way that they view incidents by shifting the perspective from a linear cause and effect approach to anticipating what may occur in the future. In addition to managing activity-based risk, predictive analytics can be used to prioritize and manage behavior-based risk. Predictive analytics allows us to move from a "rearview mirror" issues management approach to proactive behaviors that find and control weaknesses before they become real problems. When faced with a problem, traditional organizations will respond by reacting and repairing the immediate issue, an approach that generally assumes linear relationships and is limited in effectiveness. Using predictive analytics and the concepts of resilience allows us to go beyond a simple cause and effect viewpoint to a mind-set that finds, corrects, and learns from weaknesses before they become larger problems. Using predictive analytics as a tool, organizations have the opportunity to determine what changes are on the horizon that might put their success at risk through a systematic process that considers work team engagement and past operating experience in high-risk work environments. Predictive analytics is one method for identifying indications of performance degradation associated with the behaviors and practices (i.e., organizational culture) of the organization. Practical guidance for assessing and improving these behaviors and practices is discussed in Chapter 5.

5 Toolkit for Assessing and Monitoring Leadership and Safety Culture

The first four chapters have illustrated that culture is influenced by leaders' actions and the values that they communicate. Leadership commitment is essential to cultivate a strong operational culture that is characterized by safe and resilient organizational performance. Fundamentally, leaders must ensure the processes and systems provide the right balance of safety, security, and quality without introducing unanticipated negative consequences. In addition to overseeing the development and deployment of traditional systems, leadership must establish mechanisms to assess, monitor, and strengthen the organization's practices and behaviors associated with safety culture. Management commitment to a strong safety culture engages staff and has the potential to create an organization that is resilient when faced with failure.

The following chapter provides a practical toolkit for assessing, monitoring, and improving organizational leadership in the context of safety culture and high reliability. Improvement is dependent on senior leadership's understanding of how culture impacts the success of the organization. To be successful, leaders must champion the process. Most importantly, senior leaders must commit to learning and adapting based on the results of the assessment to achieve long-term positive cultural change.

5.1 ORGANIZATIONAL ATTRIBUTES OF LEADERSHIP

The organization's expectation for leadership begins by articulating the attributes that clearly tie to the institution's values and beliefs. The following section describes leadership attributes that are acknowledged in the literature as influential in contributing to a safe and resilient organization. Attributes lay the foundation for expected leadership behaviors associated with a strong safety culture. These attributes (or a subset) become the basis of the assessment and form the lines of inquiry used in the cultural assessment.

- Leadership engagement and time in the workplace
    - Leaders visit the workplace frequently. The presence of leadership in the workplace creates an understanding of worker challenges and concerns and provides an opportunity to reinforce expectations for performance with staff through coaching.
    - Leaders conduct walk-throughs and personal visits to understand how work is performed and barriers to success.
    - Leaders listen to and act on real-time information by staying in close contact with frontline staff.

- Open communication
  - Information is exchanged both formally and informally across organizational units. Two-way communications networks are established between management and staff.
  - Leaders encourage people to make suggestions, raise issues, and actively participate in resolution; people are encouraged to speak openly and honestly, voicing what may not be popular.
  - Leaders value both good news and bad news.
  - Leaders are skilled at responding to questions openly and honestly.
  - Leaders respond promptly to issues and provide feedback on problem resolution.
- Continuous improvement
  - Leaders use knowledge gained from past experiences to improve future performance.
  - Leaders encourage staff to identify opportunities for improvement by discussing good catches and near misses.
  - When things do not go right, leaders consistently take the opportunity to maximize learning.
  - Leaders encourage staff to use their knowledge and experience to identify and resolve problems. Individuals are engaged in designing and implementing improvement initiatives and solving problems.
  - Leaders cultivate a critical, questioning attitude that is focused on improvement.
- Clear expectations and accountability
  - Leaders encourage the reporting of errors by recognizing and rewarding self-identification.
  - Leaders view mistakes as an opportunity to learn.
  - Leaders consistently communicate performance expectations and use recognition as an opportunity to motivate staff and positively reinforce behavior.
- Staff development
  - Leaders encourage professional and technical growth.
  - Leaders coach, mentor, and reinforce standards and positive behaviors.
- Support to accomplish work activities
  - Leaders ensure that sufficient resources have been provided so staff can do their work with distinction. Resources may include manpower, financial support, and accessibility to information and equipment.
- Decision making
  - Leaders check the understanding of a situation by collaborating with others before proceeding.

5.2 GOALS OF THE ASSESSMENT

The organizational attributes of leaders are those behaviors that senior leadership of the organization believes align with the values and beliefs of the organization. Ideally, it is this set of attributes that forms the basis for assessment. During the planning phase of the assessment, goals are developed with the attributes in mind. Assessment goals should reflect the specific needs of the organization. Example assessment goals are:

- To identify the extent of alignment between observed behaviors and collective perceptions of organizational members and leadership of the organization with respect to the specific leadership-related safety culture attributes
- To understand the strengths and weaknesses of the organization with respect to the specific leadership-related safety culture attributes
- To recommend actions and identify ongoing efforts that might foster long-term positive culture change

5.3 SELECTION OF THE ASSESSMENT TEAM

The composition of the assessment team will directly impact the quality of the evaluation. An assessment team needs to have a broad range of competencies and experience. It is also important to select a team with a diversity of thinking styles. The composition should reflect a balance of functional areas, knowledge of and experience with safety culture evaluations, and an understanding of the organization. Typically, the team should be comprised of individuals internal and external to the organization. External team members will bring outside perspective to the gathering and interpretation of data. Internal members will bring their knowledge of work activities and organizational climate. At a minimum, the team should consist of a lead, an expert in safety culture, a senior management representative, and a knowledgeable assessor.

The assessment team members must have experience and training in the specific data collection methods utilized for the assessment. Some team members may have more experience than others and the experience of the team members should be reviewed to identify gaps and training needs. For those less experienced, consider giving team members the opportunity to practice techniques and obtain feedback from an experienced evaluator prior to the start of the assessment. Mentoring opportunities should be created as much as possible as this is an opportunity to develop the future leaders of the organization.

5.4 COMMUNICATING DURING THE ASSESSMENT

Communication with the workforce during the assessment engages both leadership and the workforce in the process and contributes to organizational learning. Frequent and transparent communication is a chance for leaders to articulate their vision for the organization and connect it with the assessment results. Creating a climate of open communication may also facilitate staff's willingness to raise concerns during the assessment. Communication should be made in context of the desired behavior change of the organization.

Leadership is responsible for championing the culture sustainment and improvement process as a tool to facilitate the long-term success of the organization. Leadership accomplishes this by regularly communicating the status of the assessment and follow-on actions to the organization. Communications should be clear and succinct (i.e., clearly state an expected action, what the action/deliverable looks like, and the intended impact of the action).

Ideally, communications should seek to build on previous communication. For example, if the results of a culture assessment have been previously communicated, a follow-up communication on actions taken to address comments from the workforce shows that management/leadership listened to their suggestions and values their input.

5.5 COLLECTING DATA

Many psychological and organizational traits, such as safety culture, are not directly observable or directly measurable and must be measured indirectly through a number of observable indicators (Pedhauzer & Schmelkin, 1991). A combination of data collection methods should be used to develop a comprehensive picture of culture within the organization being evaluated. The assessment team should determine the specific techniques to be used during the planning of the assessment. Methods may include the direct observation of behaviors in the workplace, questionnaires, one-on-one interviews, focus group interviews, reviews of safety culture-related processes and documents, performance indicator monitoring and trending, and results of related evaluations.

There are two schools of thought on the sequence and timing of data collection methods and both have value. The first approach collects data in parallel. This approach provides multiple data sources from which conclusions may be drawn about the organization's culture. The second approach collects data sequentially so that one method may inform the line of inquiry of the next. When selecting methods, consider the schedule and availability of resources. Some methods are more interactive and provide richer data, but are more resource intensive. By their nature, document reviews and electronic surveys are non-interactive. Electronic surveys may be developed and document reviews performed prior to the fieldwork phase of the assessment. Focus groups, one-on-one interviews, and behavioral observations are more interactive and require dedicated resources and on-site participation.

When collecting and analyzing data, be aware of bias that can be introduced through the self-selection bias of the participants. Respondents who chose to participate and provide feedback may have strong views, which could potentially bias the results. This is especially true of the qualitative information in the form of comments from surveys, focus groups, and interviews. If participation is voluntary, staff that have no opinion, or who do not see value in participating, will either not attend or not participate, which means that results may be skewed toward the stronger opinions.

5.5.1 Performing Observations

Observations can provide valuable insights into the influence of leaders on workplace behavior. Direct observations of workplace behavior provide objective and subjective information regarding the effectiveness of existing leadership. For example, observed leadership behaviors may indicate whether supervisors are receptive to concerns and support and recognize employees for raising concerns. Observation can also be a powerful validation (or not) of leaders' espoused beliefs as revealed during interviews and focus groups. For example, teamwork may be an espoused value of leaders revealed during interviews but observation may indicate that leaders reward and value individual contributions. The following are tips for conducting observations:

- No surprises. Notify the group or individuals being observed.
- Blend in. Be unobtrusive to avoid behavior modification of the individuals being observed.
- Respect participant identity. Assure the anonymity of the participants.

Observations have limitations. Remember that observations provide qualitative information and should not be quantified and used for statistical purposes. Additionally, take care not to over-generalize from too few observations and be mindful of the bias that may be introduced into the observations by the observer and by the selection of observed participants. An observation may be made for any activity to gain insight into the safety culture of the organization, including work planning meetings, union meetings, committee meetings, employee team meetings, and management meetings. Table 5.1 is an example form that may be customized and used to capture observations to meet the needs of your assessment.

Table 5.2 provides a protocol for conducting observations.

Table 5.3 provides general guidance on what to look for when observing leadership interaction with peers and direct reports (Mack et al., 2005, p. 20).

5.5.2 Performing Surveys

Surveys are intended to provide information on individual or group perceptions and how they impact attitude and behaviors. Perceptions are the way people organize and interpret their sensory input, or what they see and hear, and call reality. Perceptions give meaning to a person's environment and help them make sense of the world. Perceptions are important because people's behaviors are based on their perception of reality. Therefore, employees' perceptions of their organizations become the basis on which they behave while at work (Erickson, 2013).

Surveys provide information on employee attitudes, opinions, and perceptions. Surveys can be collected either by questionnaire, one-on-one interviews, or focus groups. The nature of survey types ranges from purely qualitative to quantitative, and each method will provide different information. For example, the large sample size possible with questionnaires can be used to reflect the perception of the general population. One-on-one interviews and focus groups with key individuals can be used to provide deeper meaning or clarify ambiguity that was revealed by questionnaire. The methods used in the assessment must align with the goals of the assessment. Figure 5.1 displays the quantitative–qualitative spectrum of five types of interview methods.

TABLE 5.1
Example Observation Form

Description of activity:

Leaders and participants present:

Activity Descriptors	Y N NA Circle Only One	Comments
Leader Behaviors		
Did the leader share information and occasionally verify understanding?	Y N NA	
Did the activity begin on time?	Y N NA	
Did the leader listen to concerns?	Y N NA	
Were individuals treated with respect?	Y N NA	
Were inappropriate behaviors challenged?	Y N NA	
Did the leader's behavior encourage candid discussions?	Y N NA	
Did the leader seek out differing points of view?	Y N NA	
Did the leader draw out less active participants?	Y N NA	
Did the leader solicit challenges to assumptions related to the activity?	Y N NA	
Did the leader actively solicit feedback?	Y N NA	
Participant Behaviors		
Did staff appear to be prepared and knowledgeable?	Y N NA	
If there were "stand-ins," did they actively participate?	Y N NA	
Did all staff participate in discussions?	Y N NA	
Did staff exhibit a strong sense of teamwork and collaboration?	Y N NA	

5.5.2.1 Developing a Questionnaire

To produce valid and reliable results, questionnaires should be designed by experienced subject matter experts. Questionnaire responses are typically collected using a Likert-type scale where each response is given a numerical value and analyzed quantitatively. Establishing the validity and reliability of the questionnaire strengthens the data yielded from data collection, which allows for greater confidence in the interpretation of the results.

Validity refers to the degree that the questionnaire actually measures what it is designed or intended to measure. Validity includes four parts: face, content, criterion-related and construct validity. Face and content validity are important first steps

TABLE 5.2
Participant Observation Protocol

An Example Participant Observation Protocol

1. Schedule the observation with the participants and their manager in advance. Confirm with the manager the day before.
2. Arrive 15 minutes early and introduce yourself to the manager(s) or leader of the activity.
3. Provide a brief summary of the purpose and scope of the observation to the manager or leader: *The purpose of my being here is simply to observe interactions. I will be listening, observing, and making notes. Just conduct your activity as you normally would. No individual names or identities will be included in my notes.*
4. Assume an observing position that is inconspicuous and out of the way. Assume a neutral posture; avoid facial expressions (e.g., frowning, head nodding, etc.).
5. During the observation, do not make comments or otherwise interject yourself into the meeting/activity that you are observing.
6. At the conclusion of the observation, thank the manager. If he/she asks for feedback, you may provide a brief summary of what you observed.
7. Record field notes of observations immediately after the observation.

in establishing construct validity because they establish accuracy and connection among the questions asked and the variables measured.

Face validity is a judgment by the subject matter experts that the indicator really measures the construct. Face validity may be established using a team of knowledgeable staff to review the order of questions being asked and the clearness or lack of ambiguity associated with the question. Questions then can be eliminated or revised accordingly. Face validity can also be established by testing the survey with a subset of the population prior to disseminating it across the organization. Test participants review and comment on the usability, structure, readability, and content of the questionnaire. A survey has content validity if the questions fall into the area under study. This determination is made by subject matter experts in the field by mapping the results to a preexisting standard.

Criterion-related validity demonstrates that survey questions are directly comparable with other measures of the same attribute. Criterion-related validity can be accomplished in several ways. Correlation comparisons can be made with responses from preexisting surveys if available. In addition, "triangulation" can be used to check and establish validity by analyzing a question from multiple perspectives such as observations, interviews, and surveys.

Construct validity evaluates whether or not the survey question measures what it is intended to measure. This is difficult. How do you determine what staff are actually thinking? Determining construct validity is an iterative process based on experience with the tool. Construct validity may be established based on the careful review of clarifying responses to each question to determine why the respondent answered the way that they did. This process occurs as data is collected over time.

TABLE 5.3
Guidance for Conducting Observations

Category	Includes	Observers Should Note
Verbal behavior and interactions	Who speaks to whom and for how long, who initiates interaction, languages or dialects spoken, tone of voice	Gender, age, ethnicity, profession
Physical behavior and gestures	What people do, who does what, who interacts with whom, who is not interacting	How people use their bodies and voices to communicate different emotions, what people's behaviors indicate about their feelings toward one another, their social rank, or their profession
Personal space	How close people stand to one another	What people's preferences concerning personal space suggest about their relationships
Human traffic	How and how many people enter, leave, and spend time at the observation site	Where people enter and exit, how long they stay, who they are (ethnicity, age, gender), whether they are alone or accompanied
People who stand out	Identification of people who receive a lot of attention from others	These people's characteristics, what differentiates them from others, whether people consult them or they approach other people, whether they seem to be strangers or well-known by others present Note that these individuals could be good people to approach for an informal interview or to serve as key informants

Source: Adapted from Mack, N. et al., *Qualitative Research Methods: A Data Collector's Field Guide*, Research Triangle Park, NC, Family Health International, 2005. Copyright 2005 by Family Health International.

Reliability refers to the consistency or repeatability of a test or measurement. In simple terms, an individual should provide the same score if they complete the survey at two different points in time. To determine reliability, results of the questionnaire by the same groups of individuals may be compared. To best demonstrate reliability, it is recommended that Cronbach's alpha be used. Cronbach's alpha measures scale reliability by comparing the data in many half-split ways and computing the correlation coefficient for each split. A Cronbach's alpha result of greater than or equal to 0.7 is considered acceptable.

The results of a questionnaire are not valid unless they represent the surveyed population. Response rate is a commonly accepted indication of representativeness. Questionnaire response rates are best addressed during the design and data collection phases of the assessment. This can be done by pre-testing the survey, increasing the data collection period, and sending reminders throughout the data collection period. While the survey is being conducted, it is advisable to monitor response rates. Survey research expert Babbie (2007, p. 262) asserts that "a response rate of at least 50 percent is considered adequate for analysis and reporting. A response of 60 percent is good; a response rate of 70 percent is very good." Many experts agree

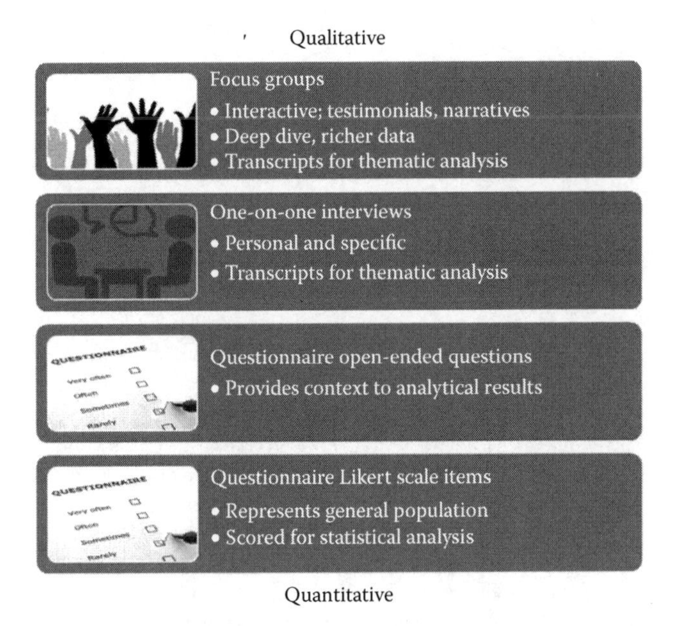

FIGURE 5.1 Interview methods: Quantitative to qualitative.

that below 50% the data should be evaluated for non-response bias (Babbie, 2007). Non-response bias is the bias that results when respondents differ in meaningful ways from non-respondents. There are many variables that could affect non-responders. For example, groups of people who fail to respond in the study could be reluctant to respond, too busy to respond, or have negative beliefs about how the organization handles survey data. Substantial differences between respondents and non-respondents make it difficult to assume representativeness across the entire population (Dillman, 1999). One method to check for non-response bias is to compare response rates across key subgroups of the target population (Groves, 2006). This may point to subgroups that could be underrepresented or justify the representativeness of the responses across the surveyed population.

In addition to the design considerations previously mentioned, the questions must clearly communicate the role or function in the organization that is being rated. For example:

- The word “I” refers to the respondent’s personal perceptions and feelings.
- The word “leader” refers to the person that has the most influence over the respondent’s day-to-day work activities.
- The word “we” refers to the entire organization.

Questionnaire items are typically placed into theoretical constructs based on behaviors that represent the construct (Crocker & Algina, 1986). A construct is an “unobservable or latent concept that the researcher can define in conceptual terms but cannot be directly measured…or measured without error. A construct can be

TABLE 5.4
Example Survey Questions Presented by Construct

Construct	Survey Question
Development	My leader promotes personal development
Safety culture	Safety is a core value for me
	Safety is a core value for my leader
	Staff and subcontractors can openly discuss issues of safety and quality
	I am empowered to ensure my own safety and that of people I work with
Accountability and recognition	My leader handles disciplinary actions fairly
	In my organization achievements are recognized and celebrated
Visibility	My leader visits my workplace routinely
	My leader visits my workplace and is involved in what's going on
	My leader understands how work is actually performed
Improvement	My workgroup anticipates what could go wrong with work activities.
	I feel encouraged to report events, concerns, and issues
	The organization responds quickly to prevent errors from recurring
	Feedback is respected by my leader regardless of rank/status of staff providing feedback
Support and trust	My leader creates an environment that is trusting and open
	My leader treats me with dignity and respect
	I trust my leader
	My leader appreciates everyone's voice regardless of weight or status
	My leader is not swayed by pressure of production when responding to issues of quality and safety
Open communications	My organization encourages openness and dialog
	I feel comfortable reporting near misses and close calls
	My leader resolves conflicts fairly
	When I make a mistake I am not afraid to report it

defined in varying degrees of specificity, ranging from quite narrow concepts to more complex or abstract concepts, such as intelligence or emotions" (Hair et al., 2006, p. 707). In the case study described in Chapter 3, seven of the survey questions were placed into three constructs of collaboration, inclusiveness, and empowerment. The strength of the relationship of questions within a construct can be measured by calculating Pearson correlation coefficient values between pairs of questions having the same dimension value. Factor analysis is recommended over Pearson's *r* when the questionnaire results produce a ratio of 10 respondents: 1 question.

Example survey questions related to leadership and safety culture are listed in Table 5.4. The subject of each of the questions (self, leader, workgroup, organization) is inserted for illustration only and should be edited depending on the design and intent of the survey.

5.5.2.2 Performing Interviews

The individual interview gives the chance to learn about the individual on a more personal level. It also offers the opportunity to ask specific questions, clarify points of interest, and elaborate on their specific comments. Depending on the design of the evaluation, interviews may occur after a questionnaire has been conducted to probe for details, examples, and deeper reasons behind the data or may be conducted as an independent collection method. Organizational assessment teams typically perform semi-structured interviews using questions that are developed from lines of inquiry. A predefined structure helps to direct the discussion so that all important aspects are covered.

To collect the most detailed and rich data from an interviewee, the interviewer uses a combination of open-ended questions and active listening techniques. Active listening techniques include:

- Paying attention by looking directly at the speaker.
- Focusing completely on the speaker.
- Taking note of cues such as body language and feelings when interviewing. It is easy to miss subtle indications that could result in missed opportunities to collect relevant data.
- Using an occasional head nod. Rather than agreeing say "I think I understand."
- Smiling and using other encouraging facial expressions.
- Using posture that is open and inviting (i.e., no crossed arms or table separating you from the interviewee).

Open questions invite the respondent to talk at length about a general subject. They ask for general information and allow the interviewee to structure his/her response. The interviewee can determine the amount and the kind of information he or she will give. Open-ended questions are good for drawing out unknown information through stories, experiences, and examples.

Initially, interview participants are placed at ease by asking questions about their general experiences as well as questions on their roles as leaders. To gain a deeper understanding as the interview progresses, the interviewer may probe into a response by asking, "Tell me more about what you were thinking?" and "How did that make you feel?"

An advantage of the interview is that the respondent can use his/her own words and expressions. It allows for greater flexibility in questioning with the possibility for follow-up questions, making it easier to get to the deeper meanings and clarify ambiguities in meaning. For example, face-to-face interviews could be an effective means for determining how leadership responds to specific worker concerns.

A difficulty with interviews is that they are not directly comparable with one another. They are also relatively time consuming, and the collective results are usually based on a limited sample, making it difficult to generalize results for the whole organization.

Table 5.5 provides an example leadership interview protocol with questions. The initial interview questions are used to facilitate dialog with the interviewee prior to probing deeper into his/her experience.

Table 5.6 provides additional example interview questions that are matrixed to the leadership attributes described in Section 5.2.

TABLE 5.5
Example Individual Interview Protocol and Questions

Phase of Interview	Protocol/ Example Questions
Introduction	Introduce yourself
	Obtain permission to conduct interviews (signed written consent)
	Point out presence of a second note taker (if applicable)
	Remind participant of confidentiality and his/her ability to stop the interview
Breaking the ice	What attracted you to work for (fill in company name)?
	When did you start?
	Describe your work experiences. What positions have you held?
Interview	Let's discuss your leadership experiences or experiences with leadership…
	If you think of a typical day for you, how would you describe your interactions? With your staff? Your manager? Your peers? What is the purpose of interaction?
	Think about an individual that you admired with respect to inspiring your engagement (i.e., when you felt attentive, absorbed, involved, and interested in what you were doing). What did they do?
	How have your interactions influenced your organization's (work team's) behavior and performance? Be specific.
	Have they changed over time? If so, what were the changes? Why did you change?
	Tell me about a time when you lead a highly engaged (or highly disengaged) work team?
	How did you know?
	What was your role in achieving and sustaining engagement?
	What specifically did you do?
	What was it like to achieve this?
Conclusion	Any final thoughts you want to share with me?
	Thank-you.

5.5.2.3 Performing Focus Group Interviews

Focus group interviews involve small groups of employees sharing their point of view about leadership within the organization. The optimal number of participants in a focus group ranges from 3 to 10. A well-conducted focus group can provide insights that are difficult to obtain through other data-gathering techniques. Focus groups rely on interaction within the group to produce insights that otherwise may not be available. As such, focus groups provide a method to collect testimonies and narratives (Denzin & Lincoln, 2000).

A skilled facilitator is essential to conducting effective focus groups. The facilitator should be impartial and have well-developed listening skills. Through body language and words, the facilitator should signal that he/she is engaged in what participants are sharing, while remaining neutral, even if they have a strong opinion. This can be done by making eye contact, nodding the head, and using phrases such as "Thank you. That is helpful."

The focus group facilitators ensure that everyone has the opportunity to speak and that one person does not dominate the conversation. One advantage of the focus

TABLE 5.6
Example Interview Questions Matrixed to Leadership Attributes

Interview Question	Leadership Attribute
Typically, how often do you interact with your staff? What is the nature of these interactions? Examples?	Leadership engagement and time in the workplace
How do you motivate staff to practice positive safety culture behaviors?	Clear expectations and accountability
What evidence is there that leaders foster desired behaviors and resolve performance issues?	Clear expectations and accountability
How do you provide coaching and feedback? Is it effective?	Staff development
How often do you see your leader in the workplace? What is the nature of the interactions? How do staff react to these interactions? Why? What value do these interactions provide? Specific examples?	Leadership engagement and time in the workplace
How are you encouraged to provide feedback, raise concerns?	Open communications
How receptive is your leader to feedback, differing opinions, and concerns? Examples?	Open communications
How does your leader demonstrate his/her commitment to improvement and solving problems?	Continuous improvement
How does your leader convey his/her expectations of your responsibilities toward safety?	Clear expectations and accountability
How does your leader balance time pressures with safety?	Support to accomplish work activities
Think about a recent period of high stress or organizational challenge. What did your leader do to engage and motivate the individual members of the team?	Support to accomplish work activities

group interview is that it allows for more people to be included in the assessment study and it can be more efficient than individual interviews by avoiding repetition.

Consideration should be given as to how well the focus group participants know each other and whether the level of familiarity will impair feedback. It is advisable to not mix employees with managers as it may stifle honest feedback.

At the beginning of the focus group session, the facilitator should present ground rules to the group to ensure that all opinions are voiced and improve the quality of data collected. The ground rules should remain visible throughout the session. Table 5.7 provides example ground rules for conducting focus groups.

Table 5.8 provides a focus group interview protocol using a specific leadership example.

5.6 DATA ANALYSIS AND INTERPRETATION

In Sections 5.2 and 5.3, the attributes and goals of the culture assessment were established. The attributes and goals formed the basis of the scope of the assessment. In Section 5.6, data were collected and organized to provide insight into the assessment goals. The next step is to analyze the data.

TABLE 5.7
Example Ground Rules for Conducting Focus Groups

Example Ground Rules for Conducting Focus Groups
Respect each other.
Conflict is okay, but avoid personal attacks.
Everyone's opinion is valid and encouraged!
Speak one at a time.
There is no right or wrong answer to questions: just ideas, experiences, and opinions, which are all valuable.
Confidentiality is assured. "What is shared in the room stays in the room."
We welcome free flow of thought and candid feedback!

TABLE 5.8
Example Focus Group Protocol

Leadership Focus Group Protocol

Introduce self and topic. Go through the ground rules

Begin with general questions to become familiar with the focus group and break the ice.

- Ask each person to tell how long they've worked for the organization, what they do, and to describe a typical work day.

Next ask focus group participants to recall and describe their experiences.

- Can you tell me about a time when working together as a team was working well?
- Give me some details, what was the problem/project?
- What did each member do?
- What was the role of the leader?
- Tell me about a time when working as a team wasn't going well?

Note: The more specific questions are intended to provide more information on:
Leadership behaviors that have impacted trust and open communications within the group.
Leadership behaviors that encouraged or inhibited identifying and reporting small failures or near misses.

During the interpretation process, the assessment team members compare the quantitative and qualitative information to the leadership attributes to better understand the basic underlying assumptions of staff. Quantitative data (numerical values) lend themselves to statistical analysis and qualitative data (words and text) lend themselves to thematic analysis. Analyzing the information collected goes beyond listing results. Similar and reinforcing data sources should be used to gain an understanding of the gaps between leadership's espoused beliefs and why that misalignment exists. The ultimate goal of analysis is to understand the organization's basic underlying assumptions that drive behaviors. The data should be objectively analyzed and integrated to develop a comprehensive understanding before reaching a conclusion.

Note: When building or analyzing quantitative and qualitative data, it is important to be aware that outlier strengths and weaknesses can be limited to specific individual departments or workgroups. This means that collecting demographic information can be important in interpreting results and assuring that conclusions are based on a representative sample of the organization.

5.6.1 Qualitative Data Analysis

The process of qualitative analysis is iterative, moving back and forth between coding, analysis, and the assessment criteria. It begins with a detailed reading of the notes and transcripts from interviews and focus groups. Once this is complete, initial codes can be generated by associating labels or tags with the text so that patterns, relationships, and themes can be recognized. The codes are given meaningful names that provide an indication of the idea or concept that underpins the theme or category. Any parts of the data that relate to a topic are coded with the appropriate label. The process of drawing conclusions begins early in the coding process (Braun & Clark, 2006).

Once all data streams have been coded, the evaluation team has sufficient information to build an overall picture of the organization. The team members should have an overview of the topics that were perceived positively or critically and whether any subgroups differed significantly from the others. At this point, the assessment team has developed an impression of the organization's performance (Braun & Clark, 2006) and can begin the search for themes.

Themes are developed by collating, nesting, and separating codes. Ideally, you will find an overarching structure or framework for the data. Try not to list themes but to explore relationships to form an overall story of your data. A good practice is to gather together all of the text passages coded for a theme. Reading all of these passages together (while also referring back to their original context for accurate interpretation) will enable a better understanding of the theme. Often it becomes clear that there is more than one theme captured by the code, and it must be partitioned. By defining/naming themes and refining them, you are moving toward an overall story. Figure 5.2 is a graphical representation of the qualitative thematic analysis performed for the case study described in Chapter 3.

5.6.2 Quantitative Data Analysis

Quantitative analysis is most commonly used with information from questionnaires, especially those using a Likert scale format.* Quantitative analysis is generally accomplished using statistical methods. Questionnaire data are typically managed by calculating a mean and a standard deviation for each question. The analysis may require more complex information such as correlations, probabilities, and skewness, looking for associations between different data, frequencies, the likelihood

* If the questionnaire sample is determined to be representative of the population, then statistical analysis is appropriate. If not, the analysis should be limited to descriptive statistics.

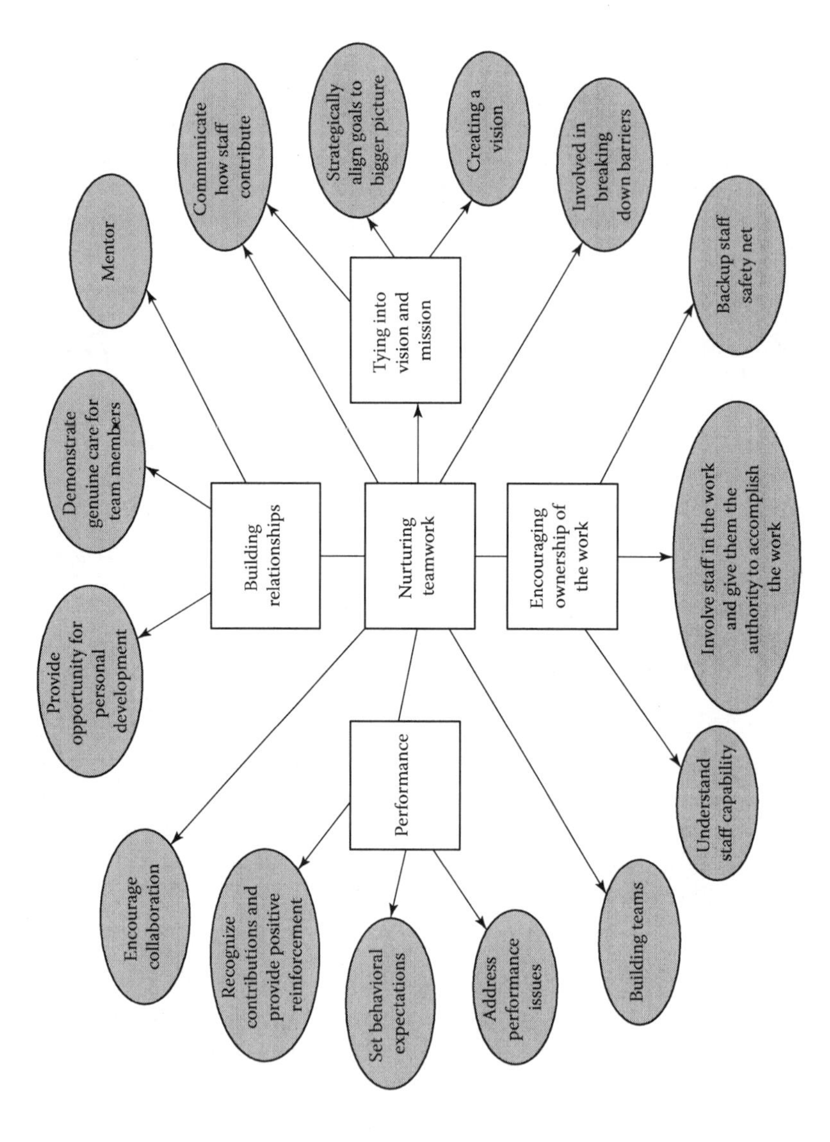

FIGURE 5.2 Thematic map used to frame results for a leadership case study.

of specific events, and outliers of data sets. Mean scores for each question may be compared between two groups using a *t*-test or multiple groups using an analysis of variance. In addition to the use of Likert-type questions, to facilitate statistical analysis, the inclusion of qualitative open-ended questions in the questionnaire is recommended. Open-ended questions invite the respondent to describe their experiences associated with a specific topic. They provide insight and context to quantitative responses.

5.7 DATA INTERPRETATION

Assessment team members interpret meaning by comparing the quantitative and qualitative information to the safety culture attributes to understand the basic underlying assumptions of the organization. The interpretation of the information must consider the context of the organization. For example, questionnaire data collected showed that 55% of respondents agreed to management visiting the workplace on a routine basis; however, 24% of staff disagreed. A review of the comments associated with the question indicated that those disagreeing appeared to be primarily staff working from home or in remote locations.

Using multiple methods (i.e., both qualitative and quantitative) for obtaining data facilitates the comparison of data to identify where there is agreement (i.e., convergence) and disagreement (i.e., divergence). When the data from the different data sources converge, the assessment team may have greater confidence in the findings. When there is divergence, more evaluation may be needed to understand why the differences exist. The assessment team is tasked to integrate the findings to answer the questions. This requires interpretation of the significance of the findings and the relationship between different findings. Conclusions should be generalizable across the organization. Always look for corroborating evidence. Don't let a single event drive conclusions. Table 5.9 provides some example results from qualitative and quantitative analysis.

Conflicting observations and findings do not necessarily indicate that the methods or analysis are invalid. While it is important to illustrate differing viewpoints, organizational assessments should look at the entire picture and conclude which of the issues, opinions, and observations characterize the organization. In some instances, differences in the comments of the assessment team may result from misunderstandings. These are resolved in the team discussion. In other cases, however, the differences of opinion between the team may reflect their different experiences in interviews and observations. Team discussion must seek an understanding of the reasons for differences of opinion. These are opportunities for the team to develop deeper insight into values and attitudes in the organization (International Atomic Energy Agency [IAEA], 2008; EFCOG/DOE, 2009).

The assessment should clarify any contradictory views and their potential meaning to prevent overemphasizing a less significant point and focusing on topics that have a relatively small impact on the overall performance. Single issues, which could severely impact safety, security, or quality, should be promptly reported to management and investigated. For example, a concern about document falsification, a

TABLE 5.9
Example Results from Mixed Method Analysis

Methods	Analysis
One-on-one interview Questionnaire	"Most people interviewed admitted that there was pressure to meet commitments but when additional time was needed to deliver a quality product it was given without cutting corners. This sentiment was supported by 88% of respondents that agreed or strongly agreed that accomplishing work safely is a priority for their manager."
One-on-one interview Observation Questionnaire	"Interviewees and observations indicated that when employees raise safety issues they are addressed immediately. Results from the survey indicate that respondents in the engineering group had statistically significant lower scores than all other organizational groups in their perception of how much emphasis management places on resolving issues."
Questionnaire One on one interview	"Sixty percent of survey respondents disagreed or strongly disagreed that quality is a priority over cost and schedule. This sentiment is supported by interviews. Several employees interviewed described examples of allowing cost and schedule to play a significant role in the decision-making process. For example, the partial incorporation of design requirements was perceived to be due to schedule pressure."

security breach, or bypassing a safety interlock needs to be reported and thoroughly evaluated.

During data interpretation, three situations may surface that merit further evaluation:

The first situation involves the extent that behaviors, practices, and values are shared within the organization. The assessment team may find, for example, a difference in the value and priority that two demographic groups place on safety. Many times, moderate variance among results is very natural considering educational, generational, and functional differences in demographics. However, if different viewpoints impact the quality of the work or prevent overall organizational development, they need to be further evaluated.

In the second situation, variance occurs because the organization's espoused values do not match practices. For example, the official policy declares that all personnel must challenge the access credentials of staff entering secured areas, although the condoned practice is to allow access without a formal check. At the least, such inconsistencies erode personnel's commitment to policies and practices. However, in some cases, the assessment team may find inconsistency simply because organizational practices are being updated and are in a process of intentional development.

The third type of situation involves variable behaviors resulting from unclear expectations. In this case, conflicting findings may reflect a lack of clear definitions and models within the organization. For example, the responsibility of workers can be emphasized across the organization, but the content of responsible behavior varies: some emphasize strict compliance with rules and written work descriptions, while others think of flexibility and an innovative mind-set.

The process of analysis and interpretation takes time. Keep questioning and testing your analysis and don't become attached to a result early in the process. Be open to alternative interpretations and seek another external review if necessary. The following is a short list of questions to consider when finalizing the assessment results:

- Do observations and questions discussed by assessment team members generally align?
- Do the staff perceptions differ with respect to the organizational subunits, roles, or other demographics?
- Does the assessment team have a collective sense for ranking findings?
- Are the conclusions based on iterations from multiple data sources and not just solitary observations?

5.8 SAFETY CULTURE IMPROVEMENT

The completed assessment provides senior leadership with the best insight into the health of the organization's culture, particularly areas that need to be improved. Sustaining and strengthening culture is difficult and requires commitment and ownership by leaders throughout the organization. For sustainable change to occur, practices must turn into habits (automatic behaviors) which when reinforced across the organization become shared assumptions. The following section describes the elements necessary to improve and sustain specific aspects of organizational culture after the assessment is complete. Figure 5.3 illustrates the process of culture assessment, sustainment, and improvement.

After the assessment is complete, organizational leadership must determine what to do with the information. Ultimately, they will decide if any additional action is necessary to strengthen and sustain critical cultural attributes. If improvement is

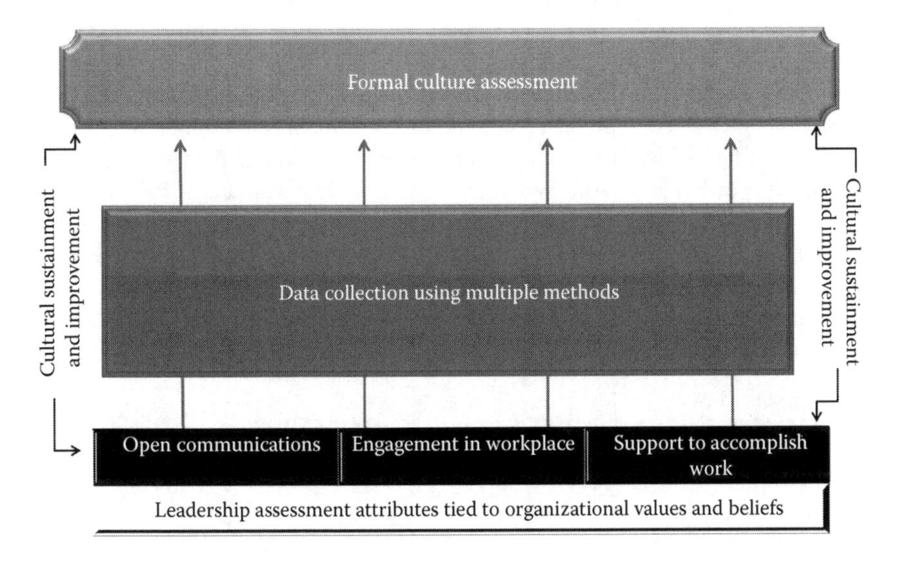

FIGURE 5.3 Culture assessment, sustainment, and improvement.

necessary, strategic outcomes and specific tactics should be documented in a plan for improvement. The plan will serve as both a compass and a road map to assess progress.

Although the importance of a strong safety culture is well accepted within high-risk industries, there is little guidance on how to take action to improve and mature the culture. There is general advice on managing organizational change but a systematic approach to designing and assessing the effectiveness of cultural interventions is lacking. Organizational change that affects behaviors and cognitive processes is acknowledged as difficult and usually unsuccessful. Burke (2011) noted that the success rate of planned change strategies persists at 30%. According to Branch and Olson (2011), to be successful in the design of cultural change strategies it is important to

- Understand safety culture within the broader context of the organization
- Have a sophisticated understanding of the perceptions and behaviors of the organization and how they relate to one another
- Recognize that assumptions, values, and beliefs may not be readily observable

Leadership must direct and manage change, including assessing the organization's readiness for change. In the simplest terms, leadership manages change by developing a strategy, communicating the strategy, empowering staff to take action, and creating short-term success that can be anchored in the organization's culture.

To be most effective, an improvement plan should involve the stakeholders within the organization. It is critical that senior leaders of the organization believe in the need for change, and agree that the entire organization should be involved in defining the change and committing to it. To be successful, management must honor the process and the design of the products must be open to the influence of all. One approach to developing improvement initiatives is to meet with all organizational stakeholders to design changes together using a facilitator. The facilitator guides the participants who are the collective decision makers. Senior leaders are present as participants. The approach has five parts: reviewing the past, mapping the present, creating an ideal future, developing a shared vision, and drawing up action plans (Griffin & Purser, 2010). When developing a change strategy, it is important to remember that the organization's existing culture is responsible for previous successes and many of the positive culture attributes can be emphasized to strengthen and overcome those culture attributes that need attention. Leveraging things that the organization is doing well supports employee engagement and facilitates ownership of the solutions to overcome weaknesses.

5.9 MONITORING AND SUSTAINING CULTURE

Once an improvement plan is set in motion, the organization can begin monitoring performance and effectiveness. Inputs to the operational culture monitoring process should emphasize behaviors rather than compliance and consider culture in terms of the organization. Data collection should integrate into existing processes and tools as much as possible and data should be organized to establish a means to monitor the status of improvement initiatives and leadership attributes that are critical to the organization's success.

5.9.1 Performance Indicators

Performance indicators should be constructed to monitor trends and provide insight into specific cultural attributes. Performance indicators inform the organization on performance, progress toward goals, and opportunities for improvement.

Although culture cannot be measured directly, there are indicators of perceptions and behaviors that can be analyzed collectively to provide an indication of potential weaknesses that could contribute to failure. The portfolio of indicators chosen should strategically fit into a framework to meet the needs of the organization and portray an evolving picture of the organization's performance. The framework typically contains both leading and lagging indicators. The complexity and number of performance indicators depends on the organization's size and structure.

Smaller events can reveal useful information about the dynamics of the organization. Leading indicators provide information to proactively identify declining performance or precursors to undesirable events. Generally, leading indicators change before the level of risk changes. For example, less severe first aid cases can be used as a leading indicator to understand behavior and safety awareness, and indicate where the organization may be at risk of experiencing a more severe injury.

Lagging indicators reflect actual performance by measuring the outcomes of activities or events. Lagging indicators can be used in conjunction with leading indicators to better understand behaviors. For example, more severe occupational injuries could indicate a careless attitude toward personal risk.

5.9.2 The Three-Step Process for Building Aggregate Performance Indicators

Having clear goals, objectives, and expectations, and an understanding of what success looks like, is vital to developing meaningful outcomes. Once those are known, an aggregate performance indicator can be designed to characterize performance and progress.

I have found the following three-step process useful for building aggregate performance indicators:

- The first step identifies the critical organizational objectives to monitor. These organizational objectives are typically associated with key actions targeting cultural improvement that are tied to an improvement plan.
- The second step defines the specific attributes associated with the desired behavior and the overall objective identified in step 1 (refer to Section 5.2).

TABLE 5.10
An Example of the Three-Step Process for Creating Aggregate Indicators

Three Steps	Results
Leadership objective:	Leadership is engaged with staff and visible in the workplace.
Leadership attributes associated with the objective:	Leaders visit the workplace frequently. Leadership understands how work is performed, staff challenges, and barriers to success. Leadership listens to and acts on real-time information.
Performance indicator(s):	Number of walk-throughs performed. Quality of "walk-throughs as conversations" as measured by focus group feedback. Leadership responsiveness to workplace concerns as measured by focus group feedback. Survey feedback and comments: • My leader visits with me in my workplace. • My leader understands how my work is performed. • My leader encourages openness and dialog.

- The third step selects indicators to monitor performance. The portfolio of indicators should strategically fit into a cultural framework that portrays an evolving picture of performance. The framework may contain both leading and lagging indicators, but ultimately the aggregate of indicators should provide information on whether objectives are met.

Table 5.10 illustrates the three-step process for building an aggregate performance indicator using leadership as an example. The three-step process can be used iteratively to build a performance framework that addresses those objectives that are critical to achieving the desired change and sustaining existing strengths.

Once the framework is complete, individual indicators can be developed to support it. Useful and effective indicators require planning to understand:

- How does the indicator fit into the overall strategy?
- What data or information is to be collected?
- How is the data or information collected?
- Who supplies the data?
- What is the collection schedule?
- What is success? Tie to a goal or standard.

Documenting your reasons for establishing the indicator and identifying responses to the considerations listed above will assure the sustainability of the indicator and provide guidance when performing the analysis. Table 5.11 contains an example performance indicator planning worksheet.

Figure 5.4 provides an example performance indicator associated with leadership engagement and visibility that is based on survey data.

TABLE 5.11
Example of a Performance Indicator Planning Worksheet

Performance Indicator: Frequency of Leadership Walk-Throughs

Summary Information. This leading indicator is associated with leadership objective: leadership is engaged with staff and visible in the workplace. The presence of leadership in the workplace creates an understanding of how work is performed, staff challenges, and barriers to success.

Reporting frequency	Information is collected and reported quarterly.
Data collection method and source	Leaders visiting staff complete a short one-line description of the interaction and annotate the results on an electronic spreadsheet that is maintained by the Human Resources organization. The spreadsheet is easily sortable and retrievable.

Description

The indicator assesses the frequency of workplace interactions that leaders have with those staff that they are directly responsible for. The purpose of capturing this information is to demonstrate that leadership is engaged with staff and visible in the workplace.

Calculation

The data are sorted by organizational subgroup and calculated as a function of the total number of staff in the subgroup.

[# of visits per quarter in the organizational function] ÷ [# of staff in the organizational function]

Target or threshold

Target of 95% of staff per quarter. This goal is based on a 20% increase from the previous year's results.

Notes

- The results will be skewed high when more than one interaction occurs with a staff member during the reporting period. The data will need to be monitored and multiple interactions acknowledged but not measured as part of the indicator.
- Does not include staff working in remote locations.
- Other indicators have been established to verify that the right type of behavior is being encouraged.

5.9.3 The Performance Index

When a small number of indicators are identified to frame a performance objective, using individual performance indicators is appropriate. Interpreting a large number of separate but related indicators provides a challenge that may lend itself to an index. Performance indexes aggregate multiple streams of performance information into one quantitative (numerical) or qualitative (stop light) conclusion. The purpose of an index is to give a quick, overall picture of performance. An index is a measure of how a variable or set of variables change over time. There is not a set formula or algorithm for generating indexes. Indexes are designed for a specific purpose and related indicators are chosen and combined to support the purpose of the index. Table 5.12 is an example of a performance index that weighs and compares nine related performance indicators to benchmark values.

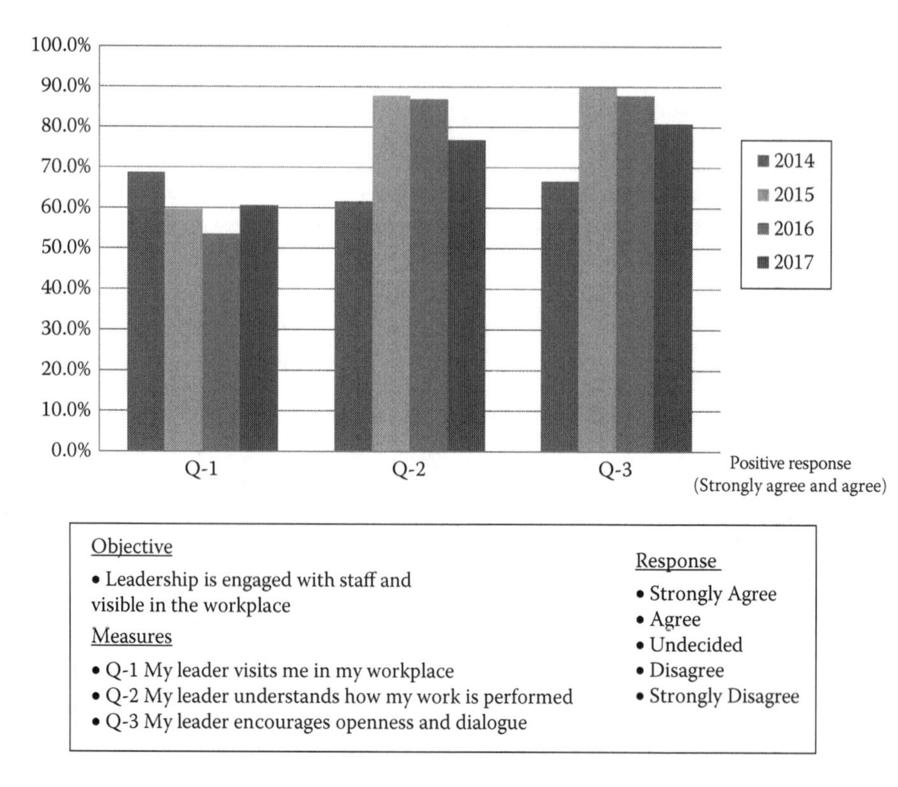

FIGURE 5.4 Performance indicator of leadership engagement and visibility.

TABLE 5.12
Example of a Performance Index

Leadership Index Indicators	Actual Data (%)	Benchmark (%)	Index	Weight (%)	Weighted Value
Safety is a core value for me (% strongly agree)	79	80	**0.99**	10	**0.10**
Safety is a core value for my immediate manager (% strongly agree)	79	80	**0.99**	10	**0.10**
At work my opinions seem to count (% strongly agree)	32	33	**0.97**	10	**0.10**
My supervisor creates an environment that is trusting and open (% strongly agree)	44	57	**0.77**	10	**0.08**
Responses to reports of hazards are timely and adequate (% strongly agree)	60	43	**1.40**	10	**0.14**

(*Continued*)

TABLE 5.12 (CONTINUED)
Example of a Performance Index

Leadership Index Indicators	Actual Data (%)	Benchmark (%)	Index	Weight (%)	Weighted Value
Safety is a core value for me (% strongly agree)	79	80	**0.99**	10	**0.10**
Safety is a core value for my immediate manager (% strongly agree)	79	80	**0.99**	10	**0.10**
At work my opinions seem to count (% strongly agree)	32	33	**0.97**	10	**0.10**
My supervisor creates an environment that is trusting and open (% strongly agree)	44	57	**0.77**	10	**0.08**
Responses to reports of hazards are timely and adequate (% strongly agree)	60	43	**1.40**	10	**0.14**
# of performance reviews completed on time	90	90	**1.00**	13	**0.13**
My supervisor understands how my work is performed (% strongly agree)	44	57	**0.77**	13	**0.10**
My supervisor visits with me in my workplace (% strongly agree)	38	38	**1.00**	13	**0.13**
% deficiencies closed within 30 days	48	57	**0.84**	10	**0.08**
Overall Leadership Index				**100**	**0.96**

5.10 EVALUATING PERFORMANCE

Indicators should be monitored for change. When analyzing a performance indicator, review the performance objective and attribute being measured. Ask yourself how the data informs the objective and associated attributes. The evaluation of performance indicators may show the need to dig deeper on a particular topic or may result in recommendations for improvement, including the type of performance indicators and the kind of information they might reveal.

Significant variance to the target or threshold may be an indication that existing conditions have changed and have influenced performance. If an indicator unexpectedly declines, it may mean that there is no data or information to track or analyze, it does not necessarily mean that performance has declined or that the indicator is no longer valid. Variance may also indicate the need for a new goal or measure. A change in performance should prompt some evaluation into what, if anything, has changed in the process that was supplying the source data or information, and whether anything has changed in the organizational culture. When the analysis is complete, determine if any additional actions are necessary to address either the indicator's usefulness or a change in performance.

For example, when monitoring an indicator that tracks issues raised by staff to support an overall objective of a questioning attitude, a significant change does not automatically indicate a shift in cultural norms. An increase or decrease should be evaluated for a change in management focus and direction regarding the expectation for raising issues. In other words, evaluate to see if organizational behaviors have changed, but don't assume they have deteriorated.

5.11 ESTABLISHING AN IMPROVEMENT TEAM

One approach to sustaining culture is to establish an "improvement team" that monitors and reports on the status of key cultural indicators to executive leadership, and recommends actions for improvement. Members of the improvement team should represent a cross section of the organization to establish a diverse experience base and viewpoints. In addition, they should have a mature understanding of how culture might impact the success of the organization.

The improvement team periodically evaluates the organization's culture, considering input from key indicators and the progress of current improvement initiatives. The evaluation includes trends relative to organizational success and status of improvements relative to identified gaps. The improvement team's analysis is briefed to executive leadership for discussion and feedback.

When reporting status to senior leadership, resist the temptation to present lots of data. The set of data should be thoughtfully constructed to provide a compelling picture of the cultural health of the organization. Data should be presented in the context of organizational meaning and relate to the leadership objectives and attributes set by senior leadership.

Specific findings should

- Connect to historical performance
- Identify factors influencing results such as the nature of operations, or significant management initiatives
- Explain significant increases or decreases

Figure 5.5 illustrates the relationship of the improvement team, data inputs, and executive leadership.

5.12 MOVING AHEAD

This final chapter provided a practical guide for assessing, monitoring, and improving the leadership aspects of your organizational culture. Appendix A contains an example leadership assessment for a chemical processing plant. Read through it as a way to review some of the insights gleaned from Chapter 5 and reflect on how you might have approached the analysis and recommendations differently. Culture is not a perfect discipline.

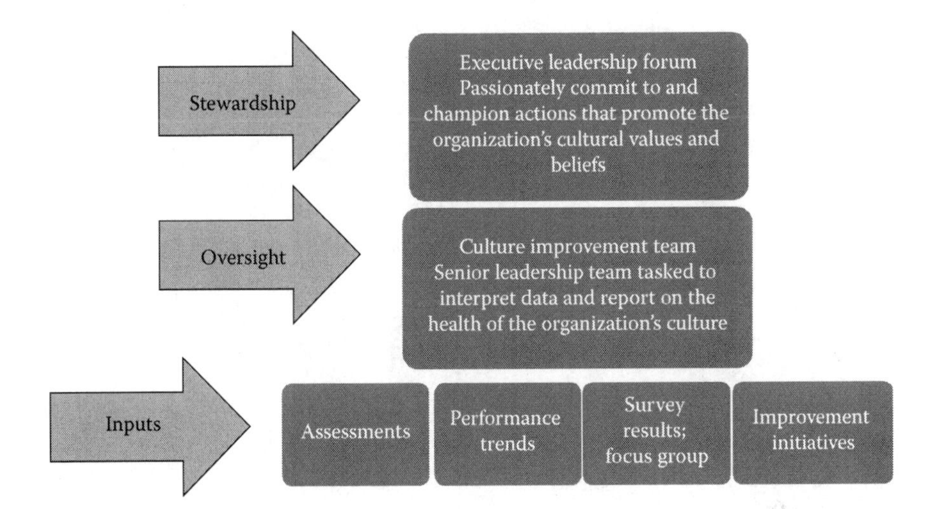

FIGURE 5.5 Relationship of the improvement team, data input, and executive leadership.

My hope is that by better understanding the factors that shape successful leadership, you will be able to reduce your organization's exposure to risk, increase your organizational resilience to failure, and at the same time develop authentic leaders of the future.

References

Apostolakis, G., & Wu, J. S. 1995. *A Structured Approach to the Assessment of the Quality Culture in Nuclear Installations.* American Nuclear Society International Topical Meeting on Safety Culture in Nuclear Installations, Vienna, Austria.

Avolio, B. J., Gardner, W. L., Walumbwa, F. O., Luthans, F., & May, D. R. 2004. Unlocking the mask: A look at the process by which authentic leaders impact follower attitudes and behaviors. *The Leadership Quarterly*, 15(6), 801–823.

Avolio, B. J., Walumbwa, F. O., & Weber, T. J. 2009. Leadership: Current theories, research, and future directions. *Annual Review of Psychology*, 60, 421–449.

Babbie, E. 2007. *The Practice of Social Research* (11th edn.). Belmont, CA: Wadsworth.

Bakker, A. B., & Demerouti, E. 2007. The job demands-resources model: State of the art. *Journal of Managerial Psychology*, 22(3), 309–328.

Bandura, A. 2000. Exercise of human agency through collective efficiency. *Current Directions in Psychological Science*, 9(3), 75–78.

Barling, J., Loughlin, C., & Kelloway, E. K. 2002. Development and test of a model linking safety-specific transformational leadership and occupational safety. *Journal of Applied Psychology*, 87(3), 488–496.

Basford, T. E., Offermann, L. R., & Wirtz, P. W. 2012. Considering the source: The impact of senior management and immediate supervisor support on employee intent to stay. *Journal of Leadership and Organizational Studies*, 19(2), 200–212.

Bass, B. M., 2008. *The Bass Handbook of Leadership.* New York: Free Press.

Bass, B. M., & Avolio, B. J. 1990. *Manual for the Multifactor Leadership Questionnaire.* Palo Alto, CA: Consulting Psychologists Press.

Bjerkan, A. M. 2010. Health, environment, safety culture and climate-analyzing the relationships to occupational accidents. *Journal of Risk Research*, 13(4), 445–477.

Bland, J. M., & Altman, D. G. 1997. Statistics notes: Cronbach's alpha. *BMJ*, 314(7080), 572.

Boin, A., & Schulman, P. 2008. Assessing NASA's safety culture: The limits and possibilities of high-reliability theory. *Public Administration Review*, 68(6), 1050–1062.

Branch, K., & Olson, J. 2011 December. Review of the literature pertinent to the evaluation of safety culture interventions technical letter report PNNL-20983. Retrieved from www.nrc.gov/docs/ML1302/ML13023A054.pdf

Braun, V., & Clarke, V. 2006. Using thematic analysis in psychology. *Qualitative Research in Psychology*, 3(2), 93.

Broughton, E. 2005. The Bhopal disaster and its aftermath: A review. *Environmental Health*, 4(6). Published online May 10, 2005. Retrieved from www.ncbi.nlm.nih.gov/pmc/articles/PMC1142333/

Browning, J. B. 1993. Union carbide: Disaster at Bhopal. Jackson Browning Report. Retrieved from www.bhopal.com/~/media/Files/Bhopal/browning.pdf

Bryant, M., & Hunter, T. 2010 September/October. BP and public issues (mis)management. *Ivey Business Journal.* Retrieved from http://iveybusinessjournal.com/publication/bp-and-public-issues-mismanagement/

Burke, W. W. 2011. A perspective on the field of organization development and change: The Zeigarnik effect. *The Journal of Applied Behavioral Science*, 47(2), 143–167.

Burns, J. 1978. *Leadership.* New York: Harper Perennial.

Caldwell, C., & Larmey, C. 2012. Preventing catastrophic incidents by predicting where they are most likely to occur and why. Presentation at Predictive Analytics World Government 2012, Washington, DC. Retrieved from www.predictiveanalyticsworld.com/gov/2012/agenda.php#day2-1125b

Caldwell, C., Matzke, B., & Larmey, C. 2017. *Predicting Accidents in a Research and Development Environment.* Manuscript in preparation.

Cassels, J. 1993. *The Uncertain Promise of Law: Lessons from Bhopal.* Toronto, ON: University of Toronto Press Incorporated.

Cheyne, A., Cox, S., Oliver, A., & Tomas, J. M. 1998. Modelling safety climate in the prediction of levels of safety activity. *Work and Stress*, 12(3), 225–271.

Christian, M., Bradley, J., Wallace, C., & Burke, M. 2009. Workplace safety: A meta analysis of the roles of person and situation factors. *Journal of Applied Psychology*, 94, 1103–1127.

Cooper, M. 2000. Towards a model of safety culture. *Safety Science*, 36(2), 111–136.

Coutu, D. L. 2002. How resilience works. *Harvard Business Review*, 80(5), 46–56.

Coyle, I., Sleeman, S., & Adams, N. 1995. Safety climate. *Journal of Safety Research*, 26(4), 247–254.

Creswell, J. W., & Plano Clark, V. L. 2011. *Designing and Conducting Mixed Methods Research* (2nd edn.). Thousand Oaks, CA: Sage.

Crocker, L., & Algina, J. 1986. *Introduction to Classical and Modern Test Theory.* New York: Holt, Rinehart & Winston.

Deal, T. E., & Kennedy, A. A. 1982. *Corporate Cultures: The Rites and Rituals of Corporate Life.* Reading, MA: Addison-Wesley.

Dekker, S. 2006. *The Field Guide to Understanding Human Error.* Burlington, ON: Ashgate Publishing.

Dekker, S. 2011. *Drift into Failure: From Hunting Broken Components to Understanding Complex Systems.* Farnham, Surrey: Ashgate Publishing.

Demerouti, E., Bakker, A. B., Nachreiner, F., & Schaufeli, W. B. 2001. The job demands: Resources model of burnout. *Journal of Applied Psychology*, 86(3), 499–512.

Denzin, N. K., and Lincoln Y. S. 2000. *Handbook of Qualitative Research.* Thousand Oaks, CA: Sage.

Deepwater Horizon Study Group (DHSG). 2011. Final report on the investigation of the Macondo Well blowout. Retrieved from http://ccrm.berkeley.edu/pdfs_papers/bea_pdfs/dhsgfinalreport-march2011-tag.pdf

Department of Energy. 2011 September 29. Integrated Safety Management System Guide (DOE G 450.4-1C, attachment 10). Retrieved from USDOE website: www.directives.doe.gov/directives/0450.4-EGuide-1c/view

Diet Report. 2012. *The National DIET of Japan: The Fukushima Nuclear Accident Independent Investigation Commission.* Retrieved from https://www.nirs.org/wp-content/uploads/fukushima/naiic_report.pdf

Dillman, D. A. 1999. *Mail and Internet Surveys: The Tailored Design Method* (2nd edn.). New York: Wiley.

Driscoll, J. 1978. Trust and decision making as predictors of satisfaction. *Academy of Management Journal*, 21(1), 44–56.

Eberly, M., Johnson, M., Hernandez, M., & Avolio, B. 2013. An integrative process model of leadership: Examining loci, mechanisms and event cycles. *American Psychologist*, 68(6), 427–443.

Edmondson, A. C. 1999. Psychological safety and learning behavior in work teams. *Administrative Science Quarterly*, 44(2), 350–383.

Edmondson, A. 2012. *Teaming.* San Francisco, CA: Jossey-Bass.

EFCOG/DOE. 2009. Safety culture task team pilot. Retrieved from http://efcog.org/wpcontent/uploads/Wgs/Safety%20Working%20Group/_Integrated%20Safety%20Management%20Subgroup/_Safety%20Culture%20HRO/Workgroup%20Documentation/_Archive/2007-2009%20Safety%20Culture%20Initiative%20Products/safety_culture_assessment_012309_final.pdf

Eid, J., Mearns, K., Larsson, G., Christian, J., Laberg, J. C., & Johnsen, B. H. 2012. Leadership, psychological capital and safety research: Conceptual issues and future research questions. *Safety Science, 50*(1), 55–61.

Erickson, J. 2013. Perception surveys: Their importance and role in safety performance. Retrieved September 26, 2014, from Safety Cary: www.predictivesolutions.com/safetycary/perception-surveys-their-importance-and-role-in-safety-performance/

Faulkner, K. 2010 May 14. BP boss implies oil slick is nothing more than a drop in the ocean. *Daily Mail.* Retrieved from www.dailymail.co.uk/news/article-1278279/Gulf-Mexico-oil-spill-BP-boss-Tony-Hayward-tries-downplay-disaster.html

Frankel, A. S., Leonard, M. W., & Denham, C. R. 2006. Fair and just culture, team behaviour, and leadership engagement: The tools to achieve high reliability. *Health Services Research*, 41(4), 1690–1709.

GAIN Working Group. 2004 September. *A Roadmap to a Just Culture: Enhancing the Safety Environment* (1st edn.). Retrieved from http://flightsafety.org/files/just_culture.pdf

Gardner, W., Avolio, B., Luthans, F., May, D., & Walumbwa, F. 2005. Can you see the real me? A self-based model of authentic leader and follower development. *The Leadership Quarterly*, 16(3), 343–372.

Gephart, R. 1988. *Ethnostatistics: Qualitative Foundations for Quantitative Research* (Qualitative research methods). Newbury Park, CA: Sage.

Gephart, R. 1993. The textual approach: Risk and blame in disaster sensemaking. *Academy of Management Journal*, 36(6), 1465–1514.

Gephart, R. 2006. Ethnostatistics and organizational research methodologies. *Organizational Research Methods*, 9(4), 417–431.

Gershon, R., Karkashian, C., Grosch, J., Murphy, L., Escamilla-Cejudo, A., Flanagan, P., Bernacki, E., Kasting, C., and Martin, L. 2000. Hospital safety climate and its relationship with safe work practices and workplace exposure incidents. *American Journal of Infection Control*, 28(3), 211–221.

Gittell, J. H., Cameron, K. S., Lim, S., & Rivas, V. 2006. Relationships, layoffs, and organizational resilience. *The Journal of Applied Behavioral Science*, 42(3), 300–329.

Governor's Independent Investigation Panel. 2011 May. Upper Big Branch. The April 5, 2010, explosion: A failure of basic coal mine safety practices. Report to the Governor retrieved from www.nttc.edu/ubb/

Griffin, T. J., & Purser, R. E. 2010. Large group interventions: Whole system approaches to organizational change. Chapter for publication in Cummings, T. (Ed). *The OD Handbook.* Thousand Oaks, CA: Sage.

Groves, R. M. 2006. Nonresponse rates and nonresponse bias in household surveys. *Public Opinion Quarterly*, 70(6), 46–75.

Guest, G., Narme, E., & Mitchell, M. 2013. *Collecting Qualitative Data: A Field Manual for Applied Research.* Thousand Oaks, CA: Sage.

Guldenmund, F. 2000. The nature of safety culture: A review of theory and research. *Safety Science*, 34(3), 215–257.

Guldenmund, F. 2007. The use of questionnaires in safety culture research: An evaluation. *Safety Science*, 45(6), 723–743.

Guldenmund, F. 2010. (Mis)understanding safety culture and its relationship to safety management. *Risk Analysis*, 30(10), 1466–1480.

Gupta, J. 2002. The Bhopal gas tragedy: How could it have happened in a developed country? *Journal of Loss Prevention in Process Industries*, 15, 1–4.

Hair, J., Black, W., Babin, B., Anderson, R., & Tatham, R. 2006. *Multivariate Data Analysis* (6th edn.). Upper Saddle River, NJ: Pearson Prentice Hall.

Hale, A. 2000. Culture's confusions. *Safety Science*, 34(1–3), 1–14.

Hamel, G., & Valikangas, L. 2003. The quest for resilience. *Harvard Business Review*, 81(9), 52–63.

Hampton, J. 2009. *Fundamentals of Enterprise Risk Management: How Top Companies Assess Risk, Manage Exposure and Seize Opportunity.* New York: AMACOM.

Haukelid, K. 2008. Theories of (safety) culture revisited: An anthropological approach. *Safety Science*, 46(3), 413–426.

Hauptman, O., & Iwaki, G. 1990. *The Final Voyage of the Challenger.* Boston, MA: Harvard Business School Publication.

Health and Safety Executive (HSE). 2003. Factoring the human into safety: Translating research into practice. Benchmarking human and organizational factors in offshore safety (Research report 059). Retrieved from www.hse.gov.uk

Health and Safety Laboratory. 2006. The causes of major hazard incidents and how to improve risk control and health and safety management: A review of existing literature (HSL/2006/117). Retrieved from www.hsl.gov.uk

Hofmann, D. A., & Morgeson, F. P. 1999. Safety as a social exchange: The role of leader member exchange and perceived organizational support. *Journal of Applied Psychology*, 84(2), 286–296.

Holling, C. S. 1973. Resilience and stability of ecological systems. *Annual Review of Ecology and Systematics*, 4, 1–23.

Hollnagel, E. 2009. *The ETTO Principle: Efficiency-Thoroughness Trade-Off.* Surrey, England: Ashgate Publishing.

Hollnagel, E., Woods, D. D., & Leveson, N. (Eds.). 2007. *Resilience Engineering: Concepts and Precepts.* Burlington, VT: Ashgate Publishing.

Hopkins, A. 2010. *Failure to Learn: The BP Texas City Refinery Disaster.* Sydney, Australia: CCH.

Horne, J. F. III, & Orr, J. E. 1998. Assessing behaviors that create resilient organizations. *Employment Relations Today*, 24(4), 29–39.

Hudson, P. 1999. Safety culture: Theory and practice. In *The Human Factor in System Reliability: Is Human Performance Predictable?* December 1–2, Siena, Italy, pp. 1–11. Retrieved from http://scholar.google.com/scholar_url?hl=en&q=http://www.dtic.mil/cgi-bin/GetTRDoc%3FAD%3DADP010445&sa=X&scisig=AAGBfm3wjyTuL7MTD_XFMrTD6zywnJOTbQ&oi=scholarr

Institute of Nuclear Power Operations (INPO). 2013 April. *Pocket Guide to INPO 12-012: Traits of a Healthy Nuclear Safety Culture.* Atlanta, GA: INPO.

International Atomic Energy Agency (IAEA). 2008. SCART guidelines: Reference report for IAEA Safety Culture Assessment Review Team. Vienna, Austria: IAEA.

International Atomic Energy Agency (IAEA). 2011. Mission report: The great East Japan earthquake expert mission. IAEA international fact finding expert mission of the Fukushima Dai-Ichi NPP accident following the great East Japan earthquake and tsunami. Retrieved from www-pub.iaea.org/MTCD/meetings/PDFplus/2011/cn200/documentation/cn200_Final-Fukushima-Mission_Report.pdf

International Dimensions of Ethics Education in Science and Engineering (IDEESE). 2009. Bhopal plant disaster case study. Retrieved from the University of Massachusetts Amherst website www.umass.edu/sts/ethics/cases.html

International Nuclear Safety Advisory Group (INSAG). 1991. Safety Culture. Safety Series No. 75-INSAG-4. Vienna, Austria: International Atomic Energy Agency.

Janićijević, N. 2011. Methodological approaches in the research of organizational culture. *Economic Annals*, 56(189), 69–99.

Johnson, R. B., & Onwuegbuzie, A. J. 2004. Mixed methods research: A research paradigm whose time has come. *Educational Researcher*, 33(7), 14–26.

Johnson, S. E. 2007. The predictive validity of safety climate. *Journal of Safety Research*, 38(5), 511–552.

Jung, D., & Avolio, B. 2000. Opening the black box: An experimental investigation of the mediating factors of trust and value congruence on transformational and transactional leadership. *Journal of Organizational Behavior*, 21(8), 949–964.

Kahn, W. 1990. Psychological conditions of personal engagement and disengagement at work. *Academy of Management Journal*, 33(4), 692–724.

Kemeny, J. G. 1979. Report of the President's commission on the accident at Three Mile Island. President's commission on the accident at Three Mile Island Washington, DC.

Kendra, J. M., & Wachtendorf, T. 2003. Elements of resilience after the world trade center disaster: Reconstituting New York City's emergency operations centre. *Disasters*, 27(1), 37–53.

Kleindorfer, P., Oktem, U., Pariyani, A., & Seider, W. 2012. Assessment of catastrophe risk and potential losses in industry (article in press). *Computers and Chemical Engineering*, 47(12), 85–96.

Kuhlicke, C. 2013. Resilience: A capacity and a myth: Findings from an in-depth case study. *Disaster Management Research*, 67, 61–76.

Lampel, J., Shamsie, J., & Shapira, Z. 2009. Experiencing the improbable: Rare events and organizational learning. *Organizational Science*, 20(15), 835–845.

Lee, A. V., Vargo, J., & Seville, E. 2013. Developing a tool to measure and compare organizations' resilience. *Natural Hazards Review*, 14(1), 29–41.

Lee, T. R. 1996 October. Perceptions, attitudes and behavior: The vital elements of a safety culture. *Health Safety*, 13, 1–15.

Lengnick-Hall, C. A., & Beck, T. E. 2005. Adaptive fit versus robust transformation: How organizations respond to environmental change. *Journal of Management*, 31(5), 738–757.

Lengnick-Hall, C. A., Beck, T. E., & Lengnick-Hall, M. L. 2011. Developing a capacity for organizational resilience through strategic human resource management. *Human Resource Management Review*, 21(3), 243–255.

Luria, G. 2010. The social aspects of safety management: Trust and safety climate. *Accident Analysis and Prevention*, 42(4), 1288–1295.

Luthans, F., & Avolio, B. 2003. Authentic leadership: A positive developmental approach. In K. Cameron, J. Dutton, & R. Quinn (Eds.), *Positive Organizational Scholarship* (pp. 241–261). San Francisco, CA: Berrett-Koehler.

Luthans, F., Luthans, K., & Luthans, B. 2004. Positive psychological capital: Human and social capital. *Business Horizons*, 47(1), 45–50.

Luthans, F., Vogelgesang, G. R., & Lester, P. B. 2006. Developing psychological capital of resiliency. *Human Resources Development Review*, 5(1), 25–44.

Luthans, F., Youssef-Morgan, C., & Avolio, B. 2015. *Psychological Capital and Beyond*. New York: Oxford University Press.

Mack, N., Woodsong, C., Macqueen, K., Guest, G., & Namey, E. 2005. *Qualitative Research Methods: A Data Collector's Field Guide*. Research Triangle Park, NC: Family Health International.

Mallak, L. A. 1997. How to build a resilient organization. In Anon (Ed.), *Industrial Engineering Solutions Conference, Proceedings*. IIE. 170–177.

Mamouni Limnios, E. A., Mazzarol, T., Ghadouani, A., & Schilizzi, G. M. 2012. The resilience architecture framework: Four organizational archetypes. *European Journal of Management*, 32(1), 104–116.

McCann, J., Selsky, J., & Lee, J. 2009. Building agility, resilience and performance in turbulent environments. *People & Strategy*, 32(3), 44–51.

McCracken, G. 1988. *The Long Interview* (Qualitative research methods). Newbury Park, CA: Sage.

McDonald, N. 2006. Organisational resilience and industrial risk. In E. Hollnagel, D. D. Woods, & N. Leveson (Eds.), *Resilience Engineering: Concepts and Precepts* (pp. 155–179). Hampshire: Ashgate Publishing.

McManus, S., Seville, E., Vargo, J., & Brunsdon, D. 2008. Facilitated process for improving organizational resilience. *Natural Hazards Review*, 9(2), 81–90.

Meyer, A. D. 1982. Adapting to environmental jolts. *Administrative Science Quarterly*, 27, 515–537.

Michael, J. H., Guo, Z. G., Wiedenbeck, J. K., & Ray, C. D. 2006. Production supervisor impacts on subordinates' safety outcomes: An investigation of leader-member exchange and safety communication. *Journal of Safety Research*, 37(5), 469–477.

Miller, K., & Monge, P. 1986. Participation, satisfaction, and productivity: A meta-analytic review. *Academy of Management Journal*, 29(4), 727–753.

Mine Safety and Health Administration. 2011 December 11. Fatal underground mine explosion: April 5, 2010. Report of investigation retrieved from www.msha.gov/Fatals/2010/UBB/PerformanceCoalUBB.asp

Montefusco, P. B., & Canato, A. 2008. A "no blame" approach to organizational learning. *British Journal of Management*, 21(4), 1057–1074.

Nahrgang, J., Morgeson, F., & Hofmann, D. 2011. Safety at work: A meta-analytic investigation of the link between job resources, burnout, engagement, and safety outcomes. *Journal of Applied Psychology*, 96(1), 71–94.

NASA Safety Center. 2013. Through a new lens: Apollo, Challenger, and Columbia through the lens of NASA's safety culture five-factor model (System Failure Case Study, 7[3]). Retrieved from http://nsc.nasa.gov/SFCS/

National Commission on the BP Deepwater Horizon Oil Spill and Offshore Drilling. 2011. Deep water: The Gulf oil disaster and the future of offshore drilling. Retrieved from www.gpo.gov/fdsys/pkg/GPO-OILCOMMISSION/pdf/GPO-OILCOMMISSION.pdf

Neal, A., & Griffin, M. 2006. A study of the lagged relationships among safety climate, safety motivation, safety behavior, and accidents at the individual and group levels. *Journal of Applied Psychology*, 91(4), 946–953.

Northouse, P. G. 2013. *Leadership: Theory and Practice* (6th edn.). Thousand Oaks, CA: Sage.

Ostrom, L., Wilhelmsen, C., & Kaplan, B. 1993. Assessing safety culture. *Journal of Nuclear Safety*, 34(2), 163–172.

Otway, H. J., & Misenta, R. 1980. Some human performance paradoxes of nuclear operations. *Futures*, 12(5), 340–357.

Pedhauzer, E., & Schmelkin, L. 1991. *Measurement, Design, and Analysis: An Integrated Approach*. Hillsdale, NJ: Lawrence Erlbaum Associates.

Perrow, C. 1981. Normal accident at Three Mile Island. *Society*, 18(5), 17–26.

Perrow, C. 1984. *Normal Accidents*. New York: Basic Books

Perrow, C. 1999. *Normal Accidents: Living with High Risk Technology*. Princeton, NJ: Princeton Press.

Peters, T. J., & Waterman, R. H. 1982. *In Search of Excellence*. New York: Harper and Row.

Pidgeon, N. F. 1991. Safety culture and risk management in organizations. *Journal of Cross Cultural Psychology*, 22(1), 129–140.

Pidgeon, N., & O'Leary, M. 2000. Man-made disasters: Why technology and organizations (sometimes) fail. *Safety Science*, 34(1–3), 15–30.

Rayner, G. 2010 May 3. Tony Hayward, the BP chief who vowed to make safety and reliability his top priorities. Retrieved from www.telegraph.co.uk/news/worldnews/northamerica/7673223/Tony-Hayward-the-BP-chief-who-vowed-to-make-safety-and-reliability-his-top-priorities.html

Reason, J. 2000. Human error: Models and management. *British Medical Journal*, 320(7237), 768–770.

Reiman, T., & Rollenhagen, C. 2010. Identifying the typical biases and their significance in the current safety management approaches. 10th International Probabilistic Safety Assessment and Management Conference, June 7–11, Seattle, WA.

Reinmoeller, P., & van Baardwijk, N. 2005. The link between diversity and resilience. *MIT Sloan Management Review*, 46(4), 61–65.

Roberts, K. 1993. Cultural characteristics of reliability enhancing organizations. *Journal of Managerial Issues*, 5(2), 165–181.

Roberts, K. H., & Bea, R. 2001a. Must accidents happen? Lessons from high-reliability organizations. *Academy of Management Executive*, 15(3), 70–79.

Roberts, K. H., & Bea, R. 2001b. When systems fail. *Organizational Dynamics*, 15(3), 179–191.

Roe, E., & Schulman, P. 2011 January. A high reliability management perspective on the Deepwater Horizon spill, including research implications. Deepwater Verizon Working Group. Retrieved from http://ccrm.berkeley.edu/pdfs_papers/DHSGWorkingPapersFeb16-2011/HighReliabilityManagementPerspectiveDeepwaterHorizonSpill-ER_PRS_DHSG-Jan2011.pdf

Safina, C. 2011. The 2010 Gulf of Mexico oil well blowout: A little hindsight. *PLoS Biology*, 9(4): e1001049.

Sawacha, E., Naoum, S., & Fong, D. 1999. Factors affecting performance on construction sites. *International Journal of Project Management*, 17(5), 309–315.

Schein, E. 2004. *Organizational Culture and Leadership.* San Francisco, CA: Jossey-Bass.

Simons, D. J., & Chabris, C. F. 1999. Gorillas in our midst: Sustained inattentional blindness for dynamic events. *Perception*, 28, 1059–1074.

Sonnet, M. 2016. Employee behaviors, beliefs, and collective resilience: An exploratory study in organizational resilience capacity. PhD thesis, Fielding Graduate University.

Sorensen, J. N. 2002. Safety culture: A survey of the state-of-the-art. *Reliability Engineering and System Safety*, 76(2), 189–204.

Spreitzer, G. M., and Mishra, A. K. 1999. Giving up control without losing control: Trust and its substitutes' effects on managers' involving employees in decision making. *Group and Organization Management*, 24, 155–187.

Stephenson, A., Seville, E., Vargo, J., & Roger, D. 2010. Benchmark resilience: A study of the resilience of organizations in the Auckland region. *Resilient Organizations Research Report 2010/03b, Resilient Organizations.* Accessed at www.resorgs.org.nz on 12/12/13.

Tashakkori, A., & Teddlie, C. 2008. Introduction to mixed method and mixed model studies in the social and behavioral sciences. In V. L. Plano-Clark, & J. W. Creswell (Eds.), *The Mixed Methods Reader* (pp. 7–26). Thousand Oaks, CA: Sage.

Teoh, S., & Seif Zadeh, H. 2013. Strategic resilience management model: Complex enterprise systems upgrade implementation. In J. Lee, J. Mao, & J. Thong (Eds.), *Proceedings of the 17th Pacific Asia Conference on Information Systems (PACIS 2013)*, Illinois, United States, 18–22 July, 1–12.

Terrell, S. 2011. Mixed-methods research methodologies. *The Qualitative Report*, 17(1), 254–280. Retrieved from www.nova.edu/ssss/QR/QR17-1/terrell.pdf

The Tokyo Electric Power Company, Inc. (TEPCO). 2011. Fukushima nuclear accident analysis report (Interim Report). Retrieved from www.tepco.co.jp/en/press/corp-com/release/betu11_e/images/111202e14.pdf

Turner, B. A. 1978. *Man-Made Disasters*. London, England: Wykeham Sciences Press. University of Washington Enterprise Risk Management (UW-ERM). 2010. Annual Report. Retrieved from http://finance.uw.edu/sites/default/files/erm/2010-erm-annual-report_0.pdf

Vaughn, D. 1996. *The Challenger Launch Decision: Risky Technology, Culture and Deviance at NASA*. Chicago, IL: University of Chicago Press.

Vogus, T. J., & Sutcliffe, K. M. 2007. Organizational resilience: Towards a theory and research agenda. *2007 IEEE International Conference on Systems, Man and Cybernetics*. Montreal, Quebec, 3418–3422.

Walumbwa, F. O., Avolio, B. J., Gardner, W. L., Wernsing, T. S., & Peterson, S. J. 2008. Authentic leadership: Development and validation of a theory-based measure. *Journal of Management*, 34(1), 89–126.

Walumbwa, F., Wang, P., Wang, H., Schaubroeck, J., & Avolio, B. 2010. Psychological processes linking authentic leadership to follower behaviors. *The Leadership Quarterly*, 21(5), 901–914.

Weston, M. J. 2010 January 31. Strategies for enhancing autonomy and control over nursing practice. *OJIN: The Online Journal of Issues in Nursing*, 15(1), Manuscript 2. Retrieved from www.nursingworld.org/MainMenuCategories/ANAMarketplace/ANAPeriodicals/OJIN/TableofContents/Vol152010/No1Jan2010/Enhancing-Autonomy-and-Control-and-Practice.html

Westphal, J. 2009 Winter. Basic concepts of a just culture. *Federation Forum*. Alexandria, VA: Federation of State Boards of Physical Therapy.

Weick, K. 1987. Organizational culture as a source of high reliability. *California Management Review*, 29(2), 112–127.

Weick, K. 1988. Enacted sensemaking in crisis situations. *Journal of Management Studies*, 24(4), 305–317.

Weick, K. E. 1993. The collapse of sensemaking in organizations: The Mann Gulch disaster. *Administrative Science Quarterly*, 38, 628–652.

Weick, K. 1996 Vol. May–June. Prepare your organization to fight fires. *Harvard Business Review*, 143–148.

Weick, K., & Roberts, K. 1993. Collective mind in organizations: Heedful interrelating on flight decks. *Administrative Science Quarterly*, 38(3), 357–381.

Weick, K., & Sutcliffe, K. 2001. *Managing the Unexpected: Assuring High Performance in an Age of Complexity* (1st edn.). San Francisco, CA: Jossey-Bass.

Weick, K., & Sutcliffe, K. 2007. *Managing the Unexpected: Resilient Performance in an Age of Uncertainty* (2nd edn.). San Francisco, CA: Jossey-Bass.

Weick, K. E., Sutcliffe, K. M., & Obstfeld, D. 1999. Organizing for high reliability: Processes of collective mindfulness. In R. S. Sutton & B. M. Staw (Eds.), *Research in Organizational Behavior* (vol. 1, pp. 81–123). Stanford, CA: JAI Press.

Wiegmann, D. A., Zhang, H., & von Thaden, T. 2001. *Defining and Assessing Safety Culture in High Reliability Systems: An Annotated Bibliography*. University of Illinois Institute of Aviation Technical Report (ARL-01-12/FAA-01-4). Savoy, IL: Aviation Res. Lab.

Wildavsky, A. 1988. *Searching for Safety*. New Brunswick, NJ: Transaction Press.

Zohar, D. 2002. The effects of leadership dimensions, safety climate and assigned priorities on minor injuries in work groups. *Journal of Organizational Behavior*, 23(1), 75–92.

Zohar, D., & Luria, G. 2003. The use of supervisory practices as leverage to improve safety behavior: A cross level intervention model. *Journal of Safety Research*, 34(5), 567–577. http://editors.eol.org/eoearth/wiki/Deepwater_Horizon_oil_spill

Appendix: Example Assessment of Leadership and Safety Culture at a Chemical Facility

A.1.1 BACKGROUND

The following assessment was conducted in a chemical processing facility that employs 200 staff. Chemical plants can pose a hazardous threat to employees and residents who live near the plants and high safety standards must be met to prevent potential accidents. This assessment was commissioned by the Chief Executive Officer (CEO) of the company as the result of a series of minor accidents and injuries attributed to human error and malfunctioning equipment. The company has been growing fast over the past five years and has always been focused on quality and service. Recently, there has been an increase in complaints about product quality and delivery.

A.1.2 PURPOSE AND SCOPE

The assessment was conducted from MM-DD-YY to MM-DD-YY. The assessment included all members of the workforce employed at the chemical processing facility, but focused on the production, service, marketing, and engineering functions. The purpose of the assessment was to better understand leadership behavior and how it contributes to a safe working environment. The scope of the assessment was limited to the following lines of inquiry:

- Supervision's engagement and time spent in the processing spaces of the facility.
  - Leaders visit the workplace frequently.
- Open communication between supervisors, management, and employees.
  - Leadership listens to and acts on real-time information.
  - Leaders encourage staff to make suggestions, raise issues, and actively participate in their resolution.
- Continuous improvement of work processes.
  - Leaders cultivate a critical, questioning attitude that is focused on improvement.

- Leadership support to accomplish work activities.
    - Leadership understands how work is performed, staff challenges, and barriers to success.

The goals of the assessment were twofold:

1. To understand the strengths and weaknesses of organizational behavior with respect to the leadership lines of inquiry stated previously.
2. To recommend actions that will foster long-term improvement for any weaknesses identified.

A.1.3 ASSESSMENT TEAM

The team was chosen for their knowledge of leadership and safety culture as well as their related experience in assessing organizational behavior. The assessment team consisted of a behavioral scientist, safety culture/leadership professional, and three experienced assessors (two internal and one external from the parent company).

A.1.4 METHODS

In addition to a review of documentation, a combination of data collection methods was used to provide a more comprehensive understanding of the attitudes and behaviors of the organization with respect to engagement, communications, improvement, and support. Methods used by the assessment team included the following:

A.1.4.1 Questionnaire

A questionnaire was distributed prior to the assessment. Completion of the questionnaire was voluntary. All employees were given daily access to a computer; therefore, the questionnaire was distributed electronically and configured to preserve anonymity as much as possible. The CEO of the company sent an e-mail to all participants inviting them to complete the questionnaire. A link to the survey was embedded in the e-mail. Reminder e-mails were sent periodically to all employees. In addition, informational presentations were made in morning meetings and table tents were placed in break rooms to encourage participation.

Responses to the survey were collected on a 5-point Likert scale and were assigned a numerical value as follows:

1. Strongly disagree
2. Disagree
3. Not sure
4. Agree
5. Strongly agree

Mean scores were collected and compared between groups using a t-test or analysis of variance. Individual responses with a value of 1 or 2 were considered negative.

Those with a 4 or 5 were considered positive. The questionnaire was completed by 79 percent of the population.

A.1.4.2 Interview

Personnel interviews were conducted with senior leaders, managers, and the production, engineering, marketing, and service department staff (see Figure A.1). Interviews were scheduled in a neutral location. Interviews were conducted by members of the assessment team who were experienced in the interview methodology. Questions were developed in advance based on the lines of inquiry. Personal identifiers were not attributed to comments. A total of 45 interviews were conducted.

A.1.4.3 Observations

Behavioral observations of day-to-day plant activities were conducted by members of the assessment team. The team members evaluated activities that provided an opportunity to observe leadership interactions with peers and direct reports. The activities observed included a senior management meeting, 10 pre-job briefs, work on the chemical processing floor, and 4 staff meetings.

A.1.4.4 Review of Key Documentation

During data collection, the team reviewed a wide variety of documents including

- Employee concerns, policies, and procedures
- Issues management and corrective action procedures
- Records from management review meetings
- Performance indicators
- Recent accident investigations
- Third-party regulatory assessments

A.1.5 SUMMARY

A combination of interviews, surveys, and observations indicated the strengths and weaknesses associated with the attributes related to leadership. The assessment team found that behaviors associated with the leadership attributes of engagement, communications, continuous improvement, and support are not consistently demonstrated across the organization. The following opportunities for improvement were identified:

- Senior leaders actively solicit feedback, listen to concerns, and communicate openly. This is not consistently reflected at the mid and supervisor level.
- Sharing of information is mostly limited to senior management communications. Mid-level managers and supervisors rarely share progress toward business goals or lessons learned from incidents with the workforce.

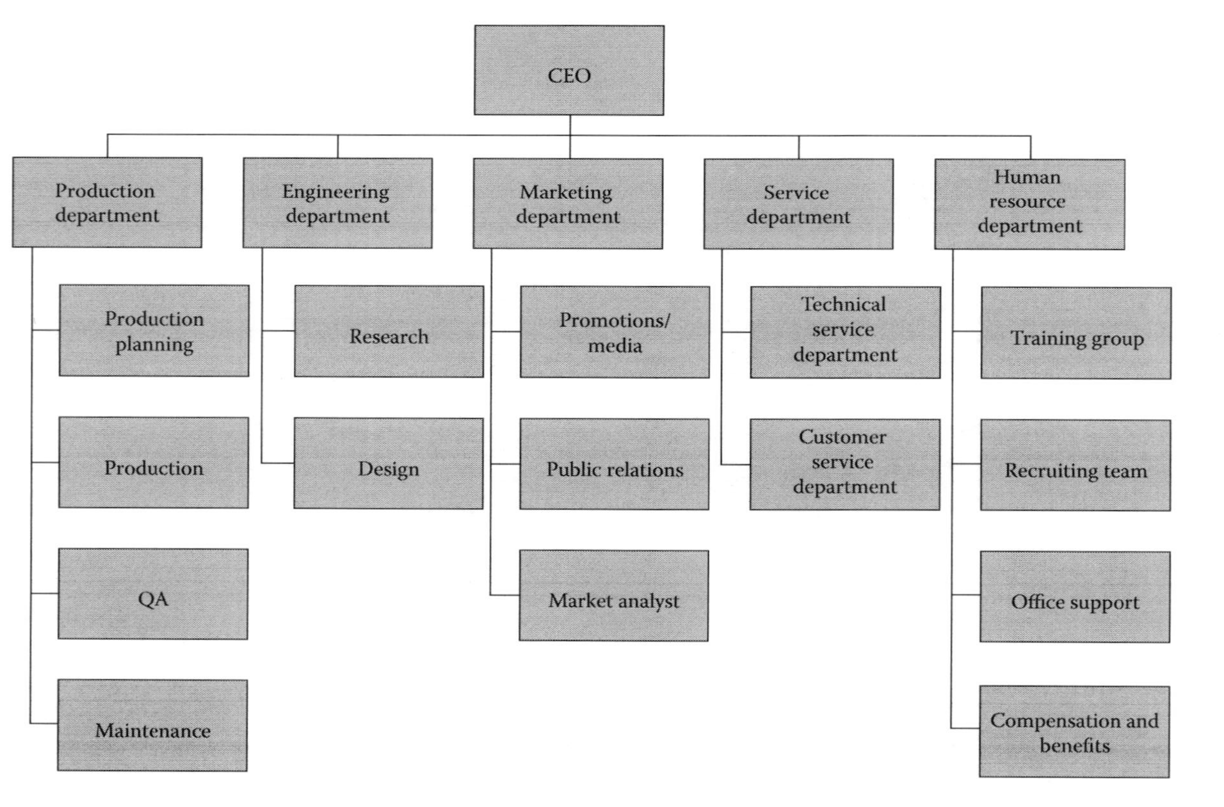

FIGURE A.1 Organization chart of the chemical processing facility

- Many employees do not feel comfortable raising minor safety concerns or stopping work.
- Some leaders are not receptive to ideas, concerns, suggestions, and differing opinions.

A.1.6 RESULTS AND DISCUSSION

Employees working in all departments at the plant appear somewhat satisfied with their jobs. The response to the question, "Overall, how satisfied are you with your job?" had a mean score of 3.7. Fifty percent of respondents scored a 4 or a 5. A statistical analysis of demographics found that there was no statistical difference in job satisfaction across the plant' s departmental functions. However, when job satisfaction scale scores were analyzed against age and length of employment, the following groups are found to be more satisfied with their overall jobs:

- Staff that have worked at the plant between 1 and 4 years.
- Staff aged 60 and older

The assessment team specifically examined four leadership areas and the results are described in the following section.

A.1.6.1 Leadership Area 1: Leadership Support to Accomplish Work Activities

Opportunity for Improvement: Management is perceived as emphasizing production over quality and safety.

Interviews with managers and supervisors indicated that most felt that there were no barriers to meeting their leadership responsibilities. Several interviewees identified barriers, such as competing priorities from additional job assignments and too many meetings of questionable value. In contrast, interviews and observations with workers indicated that most believe that their immediate supervisors are pressured to improve production numbers at the expense of safety and quality. Some workers noted that mid-level managers are not sincere and at times see their actions and behaviors as a "check the box" exercise.

A.1.6.2 Leadership Area 2: Open Communication between Supervisors, Management, and Employees

Strength: Cross-group communications is healthy and contributes to teamwork among workers.

There appears to be cohesiveness between and within workgroups at the worker level. Seventy percent of the staff responded positively in regard to open two-way communications between workgroups and only 10 percent responded negatively. This was supported by interviews and observations of pre-job briefs.

Opportunity for improvement: Some leaders are not receptive to ideas, concerns, suggestions, and differing opinions.

Survey results were mixed regarding workers' reporting concerns. Some staff in the production department indicated a hesitancy to engage with their management due to unprofessional reactions and pushback. This sentiment was supported by worker interviews that found the behaviors of their supervisors, managers, and upper levels of management changed when faced with approaching milestones and priorities. On the other hand, staff in the marketing and engineering departments responded positively to questions, indicating that workers felt they could bring up a concern without fear of retaliation and that helpful criticism was encouraged. Management in these areas are thought to encourage the reporting of concerns and appear to take concerns seriously. Interviews indicate that employees believe their management listens to them and encourages their feedback and acknowledges employee ownership and involvement in their work. Issues are generally addressed in a timely manner. In one department, a staff member interviewed spoke enthusiastically about their new manager who encourages staff to push back for clarification or to offer a differing opinion.

Issues tend to be resolved most effectively at the workgroup level. During interviews, employees said that they felt they could bring obvious safety issues to their direct supervisor. Most staff interviewed indicated that urgent safety concerns are addressed in a timely manner. Issues within the control of line managers tend to be resolved immediately. Issues that involve other organizations linger and many interviewees indicated that some issues have not been addressed.

Opportunity for improvement: Many employees do not feel comfortable raising minor safety concerns or stopping work.

Most managers interviewed revealed that they were very supportive of staff stopping work due to unsafe work conditions. Staff interviews were not as positive. Many of the staff interviewed were also able to provide examples such as inadequate work instructions, incorrect paperwork, and unexpected conditions where supervisors discouraged or challenged stop work concerns. All managers and the vast majority of staff also said that although safety was a consideration, production was also stressed. One employee stated, "The quality assurance group does not have the authority to say 'stop' when my boss is taking orders from the production manager."

A.1.6.3 Leadership Area 3: Continuous Improvement of Work Processes and the Working Environment

Strength: Senior leaders actively solicit feedback, listen to concerns, and communicate openly.

Opportunity for improvement: Many mid-level managers and supervisors feel that they do not have time to learn from minor incidents.

Opportunity for improvement: Mid-level managers and supervisors rarely share progress toward business goals or lessons learned from incidents with the workforce.

The senior managers interviewed viewed incidents and errors as a learning experience or opportunity. The senior leaders in all departments stated that they spend several hours with new employees to share expectations for performance. Lessons learned are routinely shared at leadership meetings; however, some of the mid-level managers interviewed felt that discussing incidents was a waste of time and was "bad for business" because they were "airing dirty laundry." This is an opportunity for improvement. Disseminating clear messages on injuries as learning opportunities for the organization encourages learning and reduces the probability of similar accidents. When asked about the improvement of work processes, several of the employees interviewed responded negatively.

> "The only thing that matters to my boss are the production numbers at the end of shift."
>
> "The surge in production has delayed plant maintenance. Something is going to give."

There were pockets of good practices noted. One supervisor interviewed said that he regularly chose a junior staff member to conduct a pre-job brief to give them a deeper understanding of the work activity.

A.1.6.4 Leadership Area 4: Field Presence

Opportunity for improvement: The new management observation process has improved management visibility. However, in many cases the quality of interaction is lacking.

Interviews with supervisors indicated that they are in the field daily. Managers stated a wide variation of time spent in the field ranging from a few hours to 20 hours a week. Interviews revealed that staff shared consistent sentiments. Most staff have daily interactions with their supervisor and see higher level management weekly. Some groups had limited management interaction as their work was performed in remote areas. The survey results indicated that employees have a desire for more interaction with their leadership.

First-line managers are typically trusted by workers and are most often in the work place facilitating safe work. Many workers believe that their immediate supervisors and managers are engaged and spend time in the field interacting with them. There appeared to be less confidence in the motivation of mid-level managers. Some workers indicated that levels of management above their supervisors were only fulfilling roles of demonstrated leadership because it was procedurally required.

The survey results showed that 45 percent of the employees felt managers and supervisors practice visible leadership in the field by coaching, mentoring, and reinforcing standards, while 40 percent responded negatively. The survey results were supported by field observations. The majority of manager interactions were limited to transactional topics and missed the opportunity to further engage staff. Focusing on the quality of interactions by demonstrating a genuine interest in staff and understanding their concerns about accomplishing work, tying work to the organizations mission, and sharing things that are happening will improve staff engagement and make staff feel valued.

Examples of management engagement include:

- Senior management breakfast/coffee meetings with staff. Senior management uses these meetings to share organizational developments and strategic plans, stay current on challenges, cultivate relationships, and reinforce expectations.

- Management walk-throughs using the new management observation process.
- One-on-one meetings with direct reports in the marketing and engineering departments.

A.1.7 GOOD PRACTICE

Two months ago, the CEO implemented a management observation program. Participation has been irregular and more run time is needed to determine its effectiveness. Potentially, the process will provide managers an opportunity to observe, listen, learn, and improve operations. The expectation is that managers are present to help resolve issues as they arise rather than provide oversight.

The key elements of the management observation process are:

- Management observations are required to be performed.
- The frequency for performing management observations is set by department managers.
- Managers must document their management observations.
- The intent is not an inspection or an assessment, but interaction with workers.
- Corrections are made on the spot or actions are assigned to be taken later.

Examples were given where workers believe that their immediate supervisors and managers are engaged and spend time in the field interacting with them. However, an equal number of examples were given where workers believe that management levels above their immediate supervisors show up in the field only when they have to perform their inspections/checklist activities or when something goes wrong. Many workers responded that their immediate supervisors are engaged and respond to their issues and concerns. Various comments given by both management and workers as to why mid-level managers are not in the field include:

"Managers come into the field only if there is a problem."

"Management walk-throughs are more of a 'check the box' instead of a real commitment to engaging the workers."

A.1.8 RECOMMENDATIONS

Senior leadership should use the results of this assessment as a learning opportunity. Recommendations include:

1. Reviewing the results of the assessment with all levels of management in the plant along with specific expectations for leadership.
2. Following up with an "all-hands" meeting and a pause in production to emphasize the importance of the message.

3. Developing an improvement plan and involving employees in the solution. The plan should address the following:
    a. The organization's values and mission and how each employee fits into the big picture
    b. Expectations of leaders including the relationship between production and safety
    c. Management accountability for meeting expectations
    d. The quality of management interactions with employees
    e. Organizational roles and responsibilities and the need for autonomy of oversight functions (i.e., safety and quality)

Index

I

J

K

L

M

9781138035065